"Eyal Winter's book admirably draws together the important recent work on social and individual behavior and its implications for economic behavior. He shows clearly how the more traditional rational analysis remains an important part of explanation, but is by no means adequate. His exposition is breezily informal, yet rigorous; accounts from his family join seamlessly with citations on the literature, to which he himself has made significant contributions."

—KENNETH ARROW, Nobel Laureate in Economics

Praise for *Feeling Smart*

"*Feeling Smart* puts the social back into social science. The truth is that there's a touchy feely aspect of Game Theory, and Winter shows how expressing and understanding your feelings (and those around you) will help you become a far better strategist. Be smarter or be smarting, your call."

—Barry Nalebuff, Milton Steinbach Professor, Yale School of Management, and coauthor of *The Art of Strategy*

"We are used to thinking that emotions such as anger, love, insult, and so forth are irrational. In his new book, Eyal Winter explains why these emotions are actually very rational, fulfilling important functions that usually advance the most vital interests of each of us. This is an important, enjoyable, and convincing book."

—Robert J. Aumann, Nobel Laureate in Economics

"Eyal Winter, a distinguished game theorist and behavioral economist, writes about rationality and emotion with compassion and empathy."

—Alvin Roth, Nobel Laureate in Economics

"Eyal Winter has written engagingly on the science of action and emotion; on why and how feelings make us smarter and are central to understanding rational action and interaction in processes of human betterment that are subtly inaccessible to our self-aware consciousness."

—Vernon Smith, Nobel Laureate in Economics

"Emotions and rationality are often thought of as polar opposites. But Eyal Winter—a leading game theorist and economist—shows compellingly that emotions can actually *promote* rational behavior. His book makes fascinating reading."

—Eric Maskin, Nobel Laureate in Economics

FEELING
SMART

FEELING
SMART

Why Our Emotions Are More Rational Than We Think

Eyal Winter

PublicAffairs
NEW YORK

PublicAffairs books are available at special discounts for bulk purchases
in the United States by corporations, institutions, and other organizations. For
more information, please contact the Special Markets Department at the Perseus
Books Group, 2300 Chestnut Street, Suite 200, Philadelphia, PA 19103, call
(800) 810-4145, ext. 5000, or e-mail special.markets@perseusbooks.com.

Book Design by Linda Mark
Text set in ITC Galliard 11.5 pt by the Perseus Books Group

Library of Congress Cataloging-in-Publication Data

Winter, Eyal.
 [Regashot ratsyonaliyim. English]
 Feeling smart : why our emotions are more rational than we think /
Eyal Winter.—First edition.
   pages cm
 "Published in 2012 in Hebrew in Israel, by Zmora Bitan."
 Includes bibliographical references and index.
 ISBN 978-1-61039-490-1 (hardcover : alk. paper)—
 ISBN 978-1-61039-491-8 (e-book : alk. paper)
 1. Decision making. 2. Emotions. 3. Emotions and cognition.
 4. Reason. I. Title.
 BF448.W56413 2015
 152.4—dc23

                                                      2014036706
          First Edition

          10 9 8 7 6 5 4 3 2 1

Dedicated to my wife Atalia,
who is both my emotional and my rational compass

CONTENTS

PREFACE

WHY CAN'T PEOPLE THINK MORE RATIONALLY? RELATIVE TO THE idealized image of "the thinker," evolution has seemingly bequeathed to us several flaws. How else can we explain why we are so emotional? Of what benefit is it to a person to become angry? In a world as competitive as the one we inhabit, why are we occasionally struck by a sense of humility? Why do we turn beet red, making ourselves more noticeable at precisely the moments in which we most want to bury ourselves underground out of profound shame? As long as we are on this point, why feel shame at all? Or regret? Why are we filled with a burning passion for love? And what in the world possesses us to insist on fidelity to only one lover? Or to volunteer for the most dangerous military missions? There are a plethora of actions that we would simply refuse to undertake at all if we were only to stop a moment to think about them intelligently, carefully analyze the threats versus the opportunities, and coolly calculate the net benefits. At the same time, if we were to refuse to do them, we would cease to be human beings.

Mr. Spock, a character from the television series *Star Trek*, would regularly regard his shipmates on the starship *Enterprise* with a look of forgiveness mixed with arrogance. As a native of the planet Vulcan, Spock, in marked contrast to us, acted solely out of emotionless considerations of reason and logic. Is the sense of inferiority that we feel as we watch him act calmly and coolly in the face of the grave crises that he faced on *Star Trek* justified? The truth is that if the human race had developed along the lines of the emotion-free inhabitants of Vulcan, our lives would be considerably more difficult, and in all likelihood we would not have survived at all.

Many of us tend to think of *decision making* as a process in which two separate and opposite mechanisms are engaged in a critical struggle, with the emotional and impulsive mechanism within us tempting us to choose the "wrong" thing while the rational and intellectual mechanism that we also carry inside us slowly and ploddingly promises to lead us eventually to make the right choice. This description, which was also shared by many scientists until a few decades ago, is both simplistic and wrong.

Our emotional and intellectual mechanisms work together and sustain each other. Sometimes they cannot be separated at all. In many cases a decision based on emotion or intuition may be much more efficient—and indeed better—than a decision arrived at after thorough and rigorous analysis of all the possible outcomes and implications. A study conducted at the University of California at Santa Barbara indicates that in situations in which we are moderately angry, our ability to distinguish between relevant and irrelevant claims in disputed issues is sharpened. Another study that I coauthored reveals that our inclination to become angered grows in situation in which we can benefit from anger. In other words, there is logic in emotion and often emotion in logic.

How do emotions influence our decision making? Do they hinder us or help us? What is their role in social situations? How are collective emotions formed? What are the evolutionary mechanisms that made us both thinking and emotional creatures? This book attempts to answer these questions using insights from the latest research studies published in recent years "on the seam" between emotions and rationality.

The new insights that have been obtained about the role of emotions are an outcome of a quiet revolution that has occurred over the past two decades in three important research disciplines: brain sciences, behavioral economics, and game theory. These three together have in recent years expanded our understanding of all aspects related to human behavior. If in the past emotions were studied mainly in psychology, sociology, and philosophy, while rationality was the preserve of economics and game theory, today both the study of rationality and the study of emotions are active research subjects for scholars in all those fields.

Game theory and behavioral economics, the academic fields in which I specialize, are rapidly expanding subjects within economics. Over the past two decades twelve Nobel Prizes in economics were awarded to researchers in those two fields. Their influence is felt well beyond the gates of academia. The behavioral economist Cass R. Sunstein, for example, is currently the administrator of the White House Office of Information and Regulatory Affairs in US President Barack Obama's administration. His colleague Richard H. Thaler is a member of a unit called the Behavioral Insight Team set up by British Prime Minister David Cameron in his Cabinet Office to serve as an in-house consulting board.

Although this book is not based on one and only one school of thought, it contains a personal and consistent statement. This statement can be summarized using the apparently paradoxical combination of words: "rational emotions." Research in behavioral economics and the popular literature that it has spawned, including books composed by my friends Dan Ariely[1] and Daniel Kahneman,[2] tends to concentrate on mental deviations that lead us away from rational decision making, and in some cases can harm us. In my opinion this is an overly pessimistic position. In contrast, I will try to point out how emotions serve us and further our interests, including our most material and immediate interests.

It is impossible to conduct a discussion on this subject without making use of two important research fields: game theory and the theory of evolution.

Game theory, which is essentially the study of interactive decisions, is necessary because humans are social creatures who interact with their environments. The game theoretic approach enables us to understand the roles that emotions and other behavioral characteristics have within a context of social interaction. Without it, we would be exposed to only "one side of the coin," and we would have only a partial understanding of our own behaviors.

The theory of evolution is also vital for understanding human behavior. An evolutionary claim is intended to explain how a behavioral characteristic helps (or has in the past helped) human species to survive. Like physical developments in humans and other living creatures, human behavioral developments are the results of a "package deal": a behavioral characteristic or inclination that appears to be an obstacle in one decision context is in many cases an important advantage in other decision contexts.

I have naturally emphasized the research that my research partners and I have conducted, but I have also included research results obtained by many of my colleagues and students at the Center for the Study of Rationality at the Hebrew University of Jerusalem, which I had the honor of directing for the past several years, as well as the research of many other leading scholars throughout the world. These research efforts are based on both theoretical insights and laboratory experiments, which over the past several decades have come to replace the surveys and questionnaires that had previously been the main empirical study tools of the social sciences.

My use of the term "emotions" is broader than the meaning attached to that word in common speech. I include as emotions not only concepts such as anger and worry, which are regarded by everyone as emotions, but also concepts that are typically thought of as social norms, such as fairness, equality, and magnanimity. This is not an attempt to define what is an emotion (something that I deliberately avoid doing), but instead comes from a desire to study an extensive range of phenomena that impact what might otherwise be perfectly rational thinking. The insights developed in this book are not restricted to economic decisions; they relate to a wide array of topics that include conclusions about society, politics, religion, family, sexuality, and art.

Feeling Smart is designed to enable readers who may not necessarily be up-to-date with the latest social science research to join in the fascinating discussion that is taking place on the relationship between emotions and rational behavior.

I wish to thank Benjamin Adams, who made his excellent editorial suggestions in both a rational and a sensitive manner, and my friend Ziv Hellman, who did most of the translation of this book from its original Hebrew publication in a way that no one else could do better. I owe a special debt to my research partners, my teachers, my colleagues, and my students at the Center for the Study of Rationality at the Hebrew University of Jerusalem; the intellectual interactions that I have been privileged to have with them, along with my research work, constituted the raw material for this book. Those interactions, despite being intellectual and rational, are forever also emotional for me.

INTRODUCTION

What Is Rationality?

WE BEGIN WITH SOME DEFINITIONS.

The word *rationality* is used in two different ways in nearly every spoken language. The first use relates to claims and explanations. We ascribe rationality to a given claim if it is based on a consistent internal logic and realistic assumptions.

The other common use of the word relates to decision making; this is much more complicated. To this day, economists and philosophers have not managed to agree on one direct and accepted definition. Nearly every suggested definition is either too strict (making it difficult to think of a decision that passes that definition's test of rationality) or too broad (making nearly every possible decision a rational one).

Consider a couple of examples:

DEFINITION 1: An action undertaken by an individual is rational if, given all that the individual knows, there does not exist another action that will give him or her greater material benefit (or payoff).

This appears at first glance to be a "generous" definition. Note that according to this definition the rationality of a given action is relative to the subjective knowledge that an individual has. If you purchase shares in a given company on Monday and the value of those shares falls by 50 percent on Tuesday after blaring headlines break the news that the company's CEO has been arrested and charged with financial fraud, it is still possible that you acted fully rationally according to this definition—when you made the decision to invest, you were not aware of this news item. Even if you were aware that the CEO was about to be arrested, your action would still be rational if you believed that the share price would rise anyway. As long as you believe at any given moment in time that the share price will rise, purchasing the shares is by definition a rational action at that moment.

Similarly, if you buy a rug at a Turkish bazaar at a price that is marked up by a vendor who expects his customers to bargain with him, it may be a rational choice, provided you believe that your bargaining skills are very bad and you might have lost the rug by arguing. In fact, this is a very strict definition. It restricts attention to the *material* benefit you derive from undertaking an action. According to this definition it is irrational, for example, to avoid buying shares in an expanding tobacco company because you oppose smoking. Even though this action (refraining from buying the shares) is reasonable and even admirable, it is irrational according to Definition 1 because there is another action (of which you are aware) that will increase your material benefit. The definition leaves no room for value judgments.

So, here's a broader definition:

> DEFINITION 2: An action undertaken by an individual is rational if, given all that the individual knows, there does not exist another action that will give him or her a greater amount of utility (or welfare).

Under Definition 2, an individual's decision is assessed not on the basis of the material payoff received but on the basis of "utility" or "welfare," two somewhat opaque concepts that enable this definition of rationality to relate to compensation that may be of psychological rather than material benefit. Definition 2 enables us to interpret a refusal to invest in tobacco company shares as a rational action, because the pangs of conscience that you feel may cause you to lose more, in terms of your personal welfare, than you would gain from an increase in the share price. In that case you would be better off not buying the shares.

Definition 2 is more inclusive than Definition 1. It enables altruistic actions to be considered rational because the material loss that can accompany an altruistic action may be compensated for by an uplifting mental sense of satisfaction. The problem is that the definition is too inclusive. Formally, any action may be considered rational according to Definition 2 because of the subjective nature of psychological benefits. If a man believes he will benefit psychologically from dousing himself with motor oil, then by this definition he is acting rationally if he does so. We need a definition that can describe this kind of activity as irrational. Allow me to propose a third idea, which I call the "evolutionary definition."

> DEFINITION 3: An action undertaken by an individual is rational if, given the prevailing conditions at the time the action is chosen, there does not exist another action that will give the individual a greater evolutionary advantage.

Definition 3 does judge whether an action is rational by taking into account a person's mental or material benefit from undertaking the action. A rational action must give the individual concrete (although not necessarily direct) benefit by increasing his or her evolutionary survivability. Altruistic giving, for example, may be considered rational according to Definition 3 but not for the same reason that it would be rational according to Definition 2. Definition 2 rates an altruistic action rational because of the satisfaction (the "mental compensation") gained by the giver. Definition 3 rates such an action rational because it grants the giver an evolutionary advantage.

In societies built on relations of reciprocity, an individual who assists someone else will on another occasion be assisted by others, but a selfish individual will become an outcast whose chances of survival will therefore be reduced.

The evolutionary definition depends on the environment, but in many cases it provides us with better and more accurate insight into human behavior than Definition 2. Note that suicide, which may problematically be considered rational under Definition 2, is not rational according to the evolutionary definition because suicide cannot grant an individual evolutionary advantages.

A definition of emotion will not be presented here, simply because I have yet to find a satisfactory definition from among several dozen that I have seen in all the time that I have been studying this subject. Many definitions use the expression "psychological phenomenon" but this is ultimately circular, because there is no way to define psychology without using the term emotion.

It is not surprising that we encounter difficulty in trying to define emotion. Try to imagine being asked to explain to an extraterrestrial alien the difference between a pain felt in the small toe of your left foot and the deep sorrow you feel when your closest friend dies. Or the difference between the pleasure in eating melted Belgian chocolate and what you feel when swept away in love with your spouse. We can talk about differences in the temporal duration of responses and places in the body where signals can be detected. Neurobiologists can pinpoint where different electrical stimulations are located in the brain under different emotional conditions. But these distinctions are discernible under both physical sensations and emotional feelings.

In fact, the connection between emotional and physical sensations is even stronger than we might initially think. Most of us are familiar with situations in which worry or anxiety can bring about an upset stomach or even diarrhea; on the other hand, digestive difficulties can be the source of nightmares. Yet the linkages between gut and brain don't stop there. The stomach is the only place in the body apart from the brain where extensive neurotransmitter activity takes place, especially serotonin activity (serotonin imbalance is implicated in a wide range of psychological problems, including depression). The digestive system uses serotonin for processing nutrients and passing them on to the intestines. The digestive system is the only bodily system that operates autonomously without requiring brain activity. In an incredible experiment conducted by brain scientist Michael Gershon at Columbia University in New York, part of a pig's intestines were separated from the animal's body. Food inserted in one end of the intestine passed through automatically to the other end. When a small amount of the antidepressant medication Prozac was introduced into the intestine, the digestive processing speed doubled.

Despite the surprising connections and similarities between our emotional and digestive systems, poets generally find inspiration in feelings of love or sorrow, not in digestive sensations. The reason is that the subjective experience we have in emotional situations is of its essence very different from pure bodily sensations. But even experientially we cannot find a sharp boundary line separating emotions from nonemotional bodily sensations, nor can we verbally describe the difference in a significant way. This is another reason that defining emotions is difficult.

Despite the fact that I have not found a satisfactory concise definition of emotion, there is a clear boundary between emotional behavior and rational behavior for the sake of material gain (as described in Definition 1). Behavior based on emotions is mainly understood to be automatic, while rational behavior is thought of as requiring a long and complex cognitive process that is generally slow in time. We will show, however, that these two processes often work in tandem.

There are two additional significant differences between emotions and cognitive thinking on one hand and nonemotional physiological sensations on the other hand. One of them is the fact that emotions are etched more deeply in our memories than are thoughts or even physical sensations. Many times, when we are trying to recall a film that we have seen, we discover that we have entirely forgotten the plot and even the subject of the film, but we do have a clear recollection of whether we liked the film or alternatively found it to be boring or disturbing. We can more easily retrieve a past memory of an insult or a frightening experience than of a physical pain, even if the pain was particularly intense. Memories of physical pain are often accompanied by or evoked by the emotional responses we had at the time the pain was experienced, such as anxiety or depression.

Finally, while cognitive/analytical thinking can be almost entirely controlled (in the sense that we can decide when to start or stop undertaking such thinking) and physical sensation (such as pain, for example) is almost entirely beyond our conscious control, emotions are somewhere in between. We can control our emotions to some extent and under some conditions, but not entirely. We can also evoke emotions in completely virtual situations, by external virtual stimulation (films, plays, and books), or through recollections of the past. It is no coincidence that the major film genres are categorized by emotional criteria (suspense, drama, comedy, etc.). Films offer us emotions to a much greater extent than they offer us insights.

In the following chapters we will explore the question of whether the immediacy, intensity, and flexibility of our emotional mechanisms threaten to overwhelm our rational decision-making systems, as is widely assumed, or whether instead these two mechanisms complement and assist each other.

PART I

On Anger and Commitment

1

WHAT IS THE POINT
OF GETTING ANNOYED?

Emotions as a Mechanism for Creating Commitments

IN THE AUTUMN OF 2008, AFTER PRESENTING A LECTURE AT STANFORD University, I took some time off in the cliffs overlooking the Pacific Ocean north of San Francisco. As I peered out at the ocean at twilight, the exquisite natural vista I saw filled me with a deep sense of longing. A small wedding was taking place facing the sea at the foot of the cliff on which I stood. The happy couple stood near the water's edge facing a young clergyman and a small knot of guests dressed in their finest outfits. My thoughts began to wander from the blue waters of the ocean and the red streaks of the sunset-filled sky to my wife and child, whom I had not seen in a fortnight. The sense of longing within me was accompanied by an odd combination of joy at having been privileged to have a warm and loving family, mixed with self-directed anger at the fact that I was so far from home.

Striving to amplify these feelings, I held tightly to the railing along the edge of the cliff while leaning forward to get a better glimpse of the bay and the emotion-filled wedding taking place below. Suddenly

I felt the thin railing, the only object that was preventing me from plunging directly into the abyss below, shaking. Within a fraction of a second my sentimental feelings were replaced by a powerful sense of dread that quickly propelled me away from the railing. In all likelihood that sense of dread saved my life, but it is also likely that the sense of longing that preceded it was responsible for a choice I made later: to travel less often for the sake of improving my marriage.

Emotions are a mechanism assisting us in decision making. They were formed, shaped, and developed during our evolution in order to amplify our chances of survival. If I had not felt a sense of dread when the railing I was leaning on began to shake, I would probably have continued to bend forward as it broke apart, flinging me toward my death at the bottom of the cliff. Or, if I'd fallen and somehow survived, but couldn't generate a feeling of regret, I might not have internalized the lesson learned by leaning too hard on a fragile railing. Similarly, without the capability of feeling anger toward others, we would become easy prey for exploitation and our capacity to compete over scarce resources would be weakened.

Humanity has been blessed to have, in addition to an emotional mechanism, another important mechanism assisting us in decision making—the ability to conduct rational analysis. From one point of view, it might appear that the sense of dread that I felt as the railing over the cliff began to shake was superfluous for my survival. Had I carefully calculated the extent to which the railing could bear my weight, the height of the cliff, and the full implication of my falling from the top of the cliff, I would never have leaned on the railing to begin with. But, under the circumstances, the quick reactions of my emotional mechanism were a thousand times more efficient than the slow deliberations of my rational mechanism. Rationality alone would likely have been much too slow to save my life.

In contrast to emotions such as fear, sadness, and regret, which can be defined as *autonomous emotions,* emotions such as anger, envy, hatred, and empathy are *social emotions.* They are interactive by definition. We feel anger or empathy toward others but we regret actions or situations in which we were involved. We can certainly fear others (although fear is usually induced by what another person can do to us and not by that person himself), but we do not need anyone else to feel fear. Diseases, dangers, failures, and disasters induce tremendous fear on their own.

The distinction between autonomous emotions and social emotions is especially important for understanding the concept of "rational emotions." Autonomous emotions influence our own decisions, while social emotions influence both our decisions and the decisions of others. This brings us to the most important element in the framework of emotions: their ability to create commitments, in ourselves and others. Commitment itself is one of the most important concepts in the social sciences. It is used extensively in attempts to understand economic behavior, especially with regard to bargaining theory and international relations. The Nobel Prize in economics for 2005 was awarded to Thomas Schelling mainly for his studies of commitment.

The concept of commitment is rooted in the understanding that in a conflict between two individuals, an individual who can credibly persuade his opponent that he is willing to insist on a particular outcome—even at the cost of self-harm—gains an advantage. More concretely, a seller who can credibly persuade a buyer that he has no intention of reducing the asking price of an item—even if doing so will sink the deal—is more likely to get his way. This holds true even if the buyer believes that sinking the deal will be more harmful to the seller than compromising on a lower price. In international disputes, a party to the dispute that can persuade the other party that it is willing to stick to its demands even at the price of military conflict has an advantage even, and perhaps especially, if no such armed conflict ensues.

The key rule in commitment is that the party undertaking a commitment must be truly willing to suffer the necessary sacrifice. Declarations alone cannot suffice. True commitment is difficult to counterfeit. If it were easy to fake, threats would be more common-place and no one would ever take them seriously. The fierce power that movements and nations that are fueled by religious fanaticism—such as Al Qaeda and Iran—can project is due to their ability to create credible commitments. The willingness to sacrifice welfare and even human lives for the sake of a religious idea is a potent force that gives these movements and nations significant bargaining leverage.

The barbarian Germanic tribes who crossed the Rhine River to attack the Roman Empire were able to persuade their enemies of their commitment by literally burning bridges behind them. This had the effect of proclaiming that retreat was not an option for them. For those of us who can't demonstrate commitment by setting fire to a bridge, emotions are an invaluable tool for attaining bargaining leverage in a wide range of daily conflicts. Expressing anger, for instance, shows our willingness to respond sharply to injuries or slights, even at the cost of harming ourselves, such as by starting a fistfight. If we were purely rational, we would be unable to deter our opponents so easily.

An example may illustrate the usefulness of rational emotions. Imagine yourself at an airport on the way home following a family holiday abroad. Half an hour prior to the scheduled boarding time you are informed that the flight has been canceled. You have no choice but to go to a hotel and return to the airport the following day. Now further imagine two alternative scenarios. In the first scenario you see the other airline passengers around you quietly accepting the situation and preparing to leave the terminal in an orderly manner. The boarding gate is closed and the apologetic airline offers you free transportation to the hotel of your choice. In such a scenario you are unlikely to express anger. Disappointment and frustration are more apt to be your emotions.

Now imagine a different scenario: a short time after you are informed that your flight has been canceled, you run into an acquaintance who was scheduled to fly on the same flight. She tells you that as soon as the cancellation announcement was made she went straight to the airline's representatives, made it clear to them that she had no intention of accepting the flight cancellation quietly, and demanded that an immediate solution be found enabling her to get home that same day. The result, your friend says proudly, was that the airline contacted another airline straightaway and booked her a flight home leaving in another hour.

I expect that under the second scenario your emotional state would be very different from what it would be under the first scenario. The adrenalin in your blood would rapidly rise and by the time you arrived at the airline representatives' desk to demand the same solution as your friend, you would be displaying signs of noticeable anger. In fact, you would not only be displaying signs of anger, you genuinely *would* be angry. The conscious or subconscious awareness that anger would be useful for attaining your goal would create anger within you.

The anger in the second scenario enables you to make credible threats. If in the course of speaking with airline representatives you mention an intention to sue the airline if an immediate solution is not found, your emotional state is likely to amplify the credibility of your threat. After all, a person acting solely on the basis of rational calculations would be unlikely to invest the time and money required to file such a minor lawsuit. In the first scenario, in contrast, anger would be of little help and is therefore less likely to arise.

The process creating anger in the second scenario is an astonishing interaction between the cognitive part of the brain and the limbic system that is responsible for emotional control. This process takes place in the part of the brain called the prefrontal cortex, which emerged quite late in the evolutionary development of the brain and is virtually nonexistent in other animals.

But positive emotions can show commitment, too. Love or admiration enables us to express to others our willingness to stand by their side and assist them even at a heavy price—and thus to influence their behavior toward us. Emotions need to be credible, at least at some minimal level, if they are to serve us in creating credible commitments. There are people who are able to "play-act" their emotions quite credibly, but this ability is statistically rare in the general population. If we all had perfect abilities to fake our emotions, there would never be any reason to relate seriously to the emotional responses of others and no evolutionary advantage to authentic emotional responses. Talented stage and film actors play characters in emotional situations mainly by eliciting true emotional responses from within themselves. They often do so by recalling appropriate emotional situations from their personal memories. In a sense, they are not acting but are reliving past situations. We will have more to say on credibility in later chapters.

Not every emotional reaction we have has a rational basis. In fact, most emotional reactions probably do not have a rational basis. In many cases our emotions could potentially harm us, and the ability to harness our emotions strategically, without even knowing we are doing so, is a wonderful human trait. Most of the time, using rational emotions does not require any sophistication. Indeed, children can sometimes do it more effectively than adults. A child who falls at the playground and lightly scrapes himself is more likely to cry if his mother is within eyesight. If his mother is not in the area, he is more likely to pick himself up and continue playing. He might even hold back on crying until he sees his mother. Even completely spontaneous emotions are decisively influenced by circumstances. A particular situation—for instance the audible ticking of a clock—may be exciting under some conditions (the end of a school day) but annoying under other conditions (in a doctor's waiting room). We can feel empathy or sympathy toward a given person under certain circumstances and yet feel contempt or anger toward that same person under different circumstances.

Using rational emotions and commitment as tactics is especially common in bargaining and negotiating. Emotions such as anger and insult, but also empathy, can all be identified in common negotiating situations. They influence the relative bargaining powers of the negotiators. When a labor union leader publicly states that the latest offer made by management is an embarrassment, he or she does so to improve the union's bargaining position. Such statements, however, are usually only lip service; the statement itself creates the desired sense of insult in the minds of both the labor union leadership and the rank and file workers. This has the effect of making any retreat from the commitment to turn down the offer very costly for the labor union, thus giving management an incentive to make a better offer.

People are varied in their bargaining skills. Sometimes differences in bargaining skills stem from gaps in people's abilities to create and control rational emotions or to identify them in others. There are scores of books on practical negotiating skills taught at leading business schools that call for almost entirely ignoring emotions during negotiations. I have serious reservations regarding this point of view.

In an interesting experiment, Maya Tamir of the psychology department at Hebrew University of Jerusalem induced emotional states in subjects through musical passages that they listened to.[1] Some of the musical passages had a calming effect, while others were stimulating or even irritating.

Tamir divided the subjects into two groups. One group participated individually in the task of bargaining over the division of a sum of money, and the other group participated in a collective task that required cooperation. Prior to conducting these tasks the subjects chose a musical passage to listen to. Tamir found that in the group assigned to bargain over money, the percentage of subjects who chose irritating music was significantly higher than the comparable percentage in the other group. In addition, those in the bargaining group who chose to listen to irritating musical passages achieved significantly better outcomes relative to those who chose to listen to calming music, and they walked away with much higher sums of money.

There are advantages for a negotiator who makes moderate use of emotional responses in the midst of negotiations, but the ability to control and regulate those emotions is also very important. Many negotiations break down despite the fact that a mutually beneficial agreement is available to both parties—even when both parties know that such an agreement is within reach. This usually happens when one party (or both) is stuck in a commitment (created by emotion) that is unacceptable to the other party. The chronic crises in peace negotiations between Israelis and Palestinians are a good example of this phenomenon. Emotions take over the negotiations instead of serving them, and expressions of anger and suspicion become overly convincing, defeating every attempt at arriving at an agreeable compromise.

We have so far concentrated on commitment to others. Interestingly, we use a similar mechanism to make commitments to ourselves. We often undertake actions in the present because of their effects on how we will behave in the future. A salient example of this is the purchase of a gym membership. The high cost associated with gym membership creates self-commitment to make use of its workout facilities. Another example relates to the obsessive way many of us check our incoming e-mail with such frequency that our mental concentration at work suffers. A popular computer application enables users to cut themselves off from e-mail access for a predetermined period of time. Once a user commits to an e-mail cutoff period, there is no going back; no action by the user can restore e-mail access until the full time period has run the clock. This would seem at first glance to be highly irrational: we are restricting our own freedom of action, willingly eliminating choices that we would otherwise have. But in the above examples we prefer to restrict ourselves instead of giving ourselves greater freedom, because there is a fundamental mismatch between our long-term and immediate desires (the latter often referred to as "temptations"). Our long-term desire is to attain top physical shape by visiting the gym as often as possible, but our immediate desire is often to seek the nearest good restaurant instead of working out in the gym. Self-commitment enables us to increase the cost of succumbing to immediate desire gratification before we find ourselves directly face to face with our temptations.

We often make use of self-commitment without even noticing it. If we have resolved to maintain a strict diet for the sake of reducing weight, we may studiously avoid even entering a restaurant offering an all-you-can-eat buffet, restricting ourselves to restaurants that only permit ordering à la carte. If eliminating a smoking habit is what we are striving to achieve, we will publicly announce this to our friends and acquaintances, thus attaching a painful price to slipping back into smoking: the embarrassment of public knowledge of our failure of resolve.

The phenomenon of self-commitment has a prominent place in both theoretical and empirical economic research. It forms the basis of our understanding of financial savings. Virtually every decision involving financial savings includes an aspect of self-commitment, because there we are consistently tempted to prefer consuming today rather than putting off consumption to a distant future date.

As a result, anger and shame play essential roles in financial responsibility, and even world affairs. It is possible that the recent debt crisis that brought down the economies of many countries around the world stemmed from a general lack of self-commitment, by individuals and by governments. If only these folks had been less calculating and more emotional, the story might have played out better.

2

WHY WE LOVE THOSE WHO ARE CRUEL TO US

Stockholm Syndrome and the Story of the Nazi Schoolteacher

ON AUGUST 23, 1973, A GROUP OF BURGLARS ENTERED AND commandeered a Kredibanken bank branch in Norrmalmstorg Square in Stockholm, Sweden. Over the next five days, several bank employees were held hostage in a vault by the burglars, who eventually surrendered to the authorities. What happened next was a very peculiar phenomenon. Most of the bank employees who underwent the nightmare of captivity expressed support and sympathy for the hostage takers in press interviews. Some even offered to serve as character witnesses in their defense during the subsequent trial.

About a year after these events transpired, Patricia Hearst, granddaughter of publishing magnate William Randolph Hearst, was kidnapped by a group calling itself the Symbionese Liberation Army (SLA), which had ambitions to implement a series of terrorist acts in support of radical left-wing causes, similar to the actions of the Italian Red Brigades and the Baader-Meinhof Red Army faction in Germany.

After two months in captivity, Hearst chose to join her captors, issuing a statement to the press in which she disowned her family and declared herself a member of the SLA. A short time afterward Hearst participated in a failed bank robbery along with other members of the SLA, which led to her arrest.

These two incidents, along with others, prompted psychologists and psychiatrists to identify a new psychological phenomenon termed Stockholm syndrome (or Hearst syndrome). Researchers in evolutionary psychology tend to consider Stockholm syndrome to be a behavioral phenomenon that developed in early human history. Here is the standard explanation of where it comes from. In early hunter-gatherer societies, individual tribes were competing with one another for a limited pool of food, which often led to intertribal conflict. In these situations, males would often kidnap female members of rival tribes. Natural selection favored women who successfully managed to integrate into the new tribal environment in which they found themselves: they survived and even bore the children of their captors. Women who were unable to identify emotionally with their captors usually did not survive, and if they managed to survive they often did not have offspring.

I do not regard this explanation to be fully satisfactory. First of all, Stockholm syndrome affects men as well as women. Secondly, the evolutionary explanation is too narrow and restricted relative to the wide range of expressions of the syndrome.

Stockholm syndrome is only the most extreme expression of a broader syndrome that we all, to some extent, suffer from: when we are in relationships with figures of authority, we tend to develop positive feelings toward them. People often persist in clinging to these positive feelings even in the face of injurious and unjust treatment by those in authority over them. The less opportunity people have to change their situation, the more they tend to express positive feelings toward authority figures and to blame themselves for any unjust treatment they receive at their hands. There are too many examples to list: battered women who refuse to part from their abusive husbands, unbearable bosses whose actions are inexplicably forgiven by their employees, important customers who get away with arrogant and even demeaning behavior.

I am not referring to situations in which we are fully cognizant of being in a humiliating position but stifle our anger for tactical reasons, understanding that expressing them will only be counterproductive. I am referring to cases in which we express perverse sympathy for harmful individuals or completely ignore their actions simply because they have a position of authority. A temporary boss or an unimportant customer, in contrast, will get a swift reaction from us unless the price we would pay for that reaction is too high.

In many cases, when the balance of power is especially unfavorable for us, our emotional mechanism cooperates with our cognitive mechanism to moderate our feelings of insult and anger. This is rational emotional behavior, which in proper dosage can boost our chances of survival. In extreme situations, however—as in those of battered women—that same behavioral pattern can be extremely detrimental for us. Our emotional mechanism also exaggerates the extent to which we feel gratitude toward figures of authority in return for making small and insignificant positive gestures. This can lead us to over-ascribe importance to such gestures and to develop unsubstantiated trust in the kindness and decency of the authority figure. This is the secret of success in the good cop/bad cop method of interrogating police suspects—after the bad cop has played his part and failed to elicit a confession, the good cop suddenly appears like an angel who has the suspect's best interests at heart, offering coffee or cigarettes.

I learned to appreciate the emotional power of such small gestures, even when made by particularly frightening authority figures (and perhaps especially then), from a story my father told me. In 1932 my father, Hans Winter, was the only Jewish student at the Immanuel Kant Elementary School in Königsberg, Germany. My father had a particularly vivid memory of his history teacher, Dr. Gruber, a devout Catholic, who was also an enthusiastic Nazi supporter. Gruber ignored the official Weimar Republic school program. He had his own lesson program—a virulently anti-Semitic and racist one, which taught that Germany was the cradle of human civilization while Jews were descendants of Neanderthals. He was quite aware of my father Hans's Jewishness and took great pleasure in humiliating him in front of the other schoolchildren. In one instance, for example, little Hans was called to the front of the classroom and instructed to recount the story leading up to the crucifixion of Jesus. Gruber also fully ignored the Weimar government's strict order against political rallies at schools. His lavish Nazi rallies during schooltime turned into routine practice, and when little Hans hesitantly mentioned these at home, Gruber almost lost his job. After that he would call Hans to the front of the classroom less often, but he would never take his eyes off the child.

In early February 1933, a large ceremony, orchestrated by Dr. Gruber, was held at the school to mark Hitler's appointment as chancellor of Germany. The previous administration's restrictions on political activity in schools were reversed overnight, and by eight o'clock that morning the flags and banners festooned with the swastika were ready. Fearful and indignant, little Hans decided that he could not bear to participate. He gave the flag he was carrying to the boy standing in front of him and slipped away from the parade.

Hans quickly fled from the school's parade grounds, running into the school building to hide in the bathroom. But even within the bathroom he heard someone singing the Nazi anthem from within one of the cubicles. Before he even had a chance to identify the voice of the singer, the cubicle door swung open and Hans found himself face to face with Dr. Gruber, now dressed in a starched SA uniform.

Hans made an immediate about-face and began running with all his might, with Gruber chasing right after him while at the same time trying to button the top of the fly of his trousers. "Hans Winter, halt!" roared Gruber at the top of his lungs. Hans refused even to consider this possibility, accelerating his pace even more. Hans swiftly ran out of the school grounds straight into the bustle of the city, judging that if he could reach the offices of his uncle's wheat export company, about a half-mile from the school, before Gruber could catch him, he would be safe. There was a good chance that Hans's father might be there, and if his father would see what Gruber was trying to do, he would find a way to relieve Hans of ever having to see Gruber again.

Temperatures in Königsberg in February are often far below freezing, and to Hans's misfortune, the streets were coated with a thick layer of ice that day. After only a few hectic minutes of running through the slippery city streets in the freezing cold, Hans's feet lost their grip on the ice. He went flying onto the pavement, injuring a leg in the process. As he lay flat on his back, moaning with pain, Hans could hear Gruber, gasping for air, approaching. He was certain that within seconds Gruber's heavy body would land on him, pressing his head into the ice and leaving him helpless, with no one to rescue him from the full extent of Gruber's revenge.

What happened next had a greater influence on my father's personality—for better or worse—than any other event in that fateful year in which the Nazi's took power in Germany.

Gruber gingerly approached Hans, who by this point was trying to play dead as best he could, and picked the boy up in his arms. He whispered softly: "Hans, what happened? Show me where you are hurting." After hugging Hans warmly, Gruber carefully inspected his injured foot. Hans watched Gruber warily, but with a nod of his head indicated that the pain was receding. Gruber then helped Hans stand up again, patted his head and pointed to a nearby café. After being served a cup of hot tea and a plate of chocolate cake, paid for by Gruber, Hans looked suspiciously across the table.

Gruber sat there leaning his chin on his arms, placing his head parallel to Hans's. He explained that he'd been chasing my father to make amends with him, not hurt him. "In fact I wanted to tell you that as an educator and your personal teacher I regard myself as responsible for your health and well-being in our school. No one can harm you, not a student, not a teacher, no one. Promise me that you will inform me of any attempt to hurt you." Gruber continued talking this way, stressing that now that Adolf Hitler was the leader of Germany, respect, justice, and decency would surely be the hallmarks of the new Nazi Germany. After completing this speech, he calmly turned his attention to eating the cake he had ordered for himself.

I heard my father retell this story many times. Whenever he got to describing the scene in the café his eyes welled with tears and his voice choked. Was my father reacting this way because of the general memories he had of how much he had suffered in his last year in school in Germany or perhaps because of the utter fear that had gripped him when he was being chased by Gruber, who was certainly a vile man? I do not believe it was due to either reason. I think my father reacted that way because of the kind gesture he received in the most unexpected place, at the most unexpected time, from the most unexpected person. He apparently regarded Gruber as a hero—in fact, as a sort of righteous man.

How could miserable Gruber, who exhibited a few minutes of decency, have merited this response? I never dared to ask my father this question directly, but Gruber apparently was the object of my father's empathy for years precisely because he was vile, not in spite of his usual ugly personality and behavior.

My father's emotional reaction was a moderate expression of Stockholm syndrome. Little Hans was in a situation in which he was under the authority of a teacher who made his life miserable during the very frightening time period at the beginning of the Nazi rule. The empathy that this teacher received from his pupil in exchange for a very inexpensive price is the result of a rational emotion that protected my father and enabled him to survive his last, difficult months in Germany. A particular emotion may be rational at a specific moment in time, but it can also be deeply embedded within us and survive for decades even after it has ceased to protect us.

3

EMOTIONAL IMPOSTORS, EMPATHY, AND UNCLE EZRA'S POKER FACE

THE EFFICACY OF RATIONAL EMOTIONS DEPENDS TO A GREAT EXTENT on the capacity of others to recognize such emotions and, even more importantly, to be convinced of their sincerity. No one can prevent us from feeling angry during the course of lengthy negotiations, but if we harbor hidden anger that the other party does not notice, that will give us only an ulcer, not a negotiating advantage. Artificially expressing anger that the other party will immediately spot as faked is also of no advantage to us; in fact it may work against us. Authenticity is the name of the game.

My former student, Meir Meshulam, once visited a friend of his, along with several others. As the evening prolonged, they decided to order a pizza. But the pizza failed to appear, and the young men sat around waiting, becoming increasingly frustrated. When one of the boys' father came by, he calmly asked if they had called the pizzeria to get things moving more quickly. The boys replied that they had called, but were told that the pizza was still being prepared.

The man decided to teach these young men a lesson in "how to get things done." He immediately called the pizzeria himself. His previously calm demeanor was suddenly replaced by visible anger as he shouted into the telephone, informing the person on the other end of the line that if the pizza did not arrive within five minutes, it would be the last pizza his household would ever order from that pizzeria. It might appear as if the anger he exhibited during the telephone call was entirely manufactured, especially given the calmness that he had exuded only seconds earlier, but just after ending the call he could clearly be heard saying quite angrily "those bastards!" The pizza arrived safely within the next fifteen minutes.

The point here is that we are sometimes able consciously to summon authentic emotions, even though we are doing it for strategic reasons. A few years ago Al Jazeera News interviewed me for a broadcast they were preparing on science and education in Israel. I recall that I was glad for the opportunity to create a bit of sympathy for Israel among the television station's Arab viewers. This, in fact, was my main interest in the interview.

The interview took several hours, starting with questions relating to game theory and continuing with questions regarding success of the Center for the Study of Rationality, which I directed at the time. At some stage, however, they moved into questions of a more personal nature. The interviewers were eager to learn about my family: Where were my parents born? When did they move to Israel? Was I exposed to the Palestinian history as a child? I found myself boasting that on my mother's side my family had been in Jerusalem for six generations, but I elaborated on my father's history and the way he fled from Nazi Germany. I recounted how my father and his brother were forced to leave Germany without their parents in 1933, traveling a rough and difficult route alone through Europe until they managed to get to the port of Trieste where they boarded a ship sailing to Palestine; how my father had to struggle to survive in unfamiliar surroundings that were very different from the life he knew as a child in his wealthy Jewish family in Germany. Finally I mentioned the trauma he later suffered as he received news of how the relatives he left behind were murdered in the Nazi death camps.

I had already told that same story dozens of times before to friends and relatives with hardly any emotional reaction. But as I sat across from the Al Jazeera cameras, I could not stop the tears flowing from my eyes. In retrospect I realized that I had subconsciously made myself more emotional in order to arouse sympathy among the television audience. But none of it was artificial; the grief that filled my eyes with tears was entirely authentic.

Recently, I jointly conducted a lab experiment with Meir Meshulam at the Center for the Study of Rationality.[1] For this experiment we used a device that collects data from electrodes attached to the skin, mainly relating to pulse rates and skin conductance, in order to measure the amount of emotional tension that the experiment's subjects were experiencing.

We had the subjects play a simple, two-player game called the dictator game. One of the players is given a sum of money, say $100. Both players are then told that the player holding the money has the option of sharing some of the money with the other player or keeping it all for herself—the decision is entirely hers, depending on how generous she wants to be. We were interested in the emotional reaction of subjects who were in the passive role in this game; these were the players who were connected to the skin conductivity measuring device.

Subjects were divided into three groups, with each group treated differently. We told the first group that our device could measure the anger level of the person to whom it was connected. Subjects in this first group were also told that they would be compensated if they received only a small amount from the dictator player. In addition, they were told that the amount of compensation we would give them would be proportional to the anger level we measured in them in response to a small amount offered by the dictator player. The angrier they got, they more money they would receive.

We explained to the second group that our device would measure the amount of happiness they felt as a result of receiving a generous amount of money from the dictator player. In addition, they were told that they would be rewarded if we measured them feeling happy, with the reward proportional to their happiness level. The third group was similarly incentivized to remain calm upon learning how much money they were offered by the dictator player.

Figure 1 depicts the emotional reactions of the subjects in our experiment. Emotional reactions were measured both by the skin conductivity device and by the use of questionnaires. These questionnaires contained indirect questions that have been successfully used for decades to identify emotional states.

As Figure 1 shows, subjects clearly responded to the incentives. The subjects in the first group exhibited anger in a pronounced manner when they received low offers; on the other hand, low offers did not generate much anger when players were incentivized to be happy. Interestingly, we also found that the abilities of subjects in the second group to create expressions of happiness in response to incentives were much weaker. This finding might be because the skin conductance device is less sensitive to expressions of happiness, but it might also point to greater human capacity to produce outward signs of anger rather than happiness on demand. Although anger is far less pleasant than happiness, it is much more effective in creating

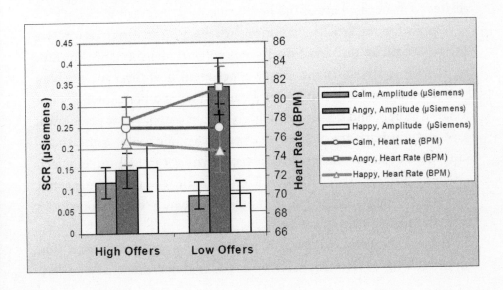

commitment in social situations. This may in turn mean that evolution selected for people whose brains were adept at expressing anger, thus making people angrier on the whole.

We all have the capacity to recognize emotional states in others. Without this capacity we would be severely limited in our ability to interact socially. Our ability to reproduce would be curtailed if we were unaware of whether or not others found us attractive. Even our sheer physical survival, which to a great extent depends on our social interactions, would be jeopardized without an ability to read emotions in others. Our capacity to pick out emotions from the faces of others apparently developed quite early in the evolution of human cognitive abilities. This processing occurs in the amygdala, which is part of the limbic (emotional) system of the brain and is located in the most interior and primal part of the human brain.

In the 1990s brain scientist Antonio Damasio and his colleagues conducted several research studies focused on subjects who had suffered injuries to their amygdalae. These subjects could easily recognize faces and could match photographs of faces with people whom they knew, but they failed utterly in recognizing facial expressions and matching those expressions to emotional states.

One of the most interesting discoveries in the field of brain science relates to a specific part of the brain called the fusiform gyrus, which is responsible for facial recognition.[2] Faces are the ultimate tool we use to broadcast our emotional states to our environment. You can try the following short and simple experiment next time you are on a public transport bus or in line at a bank: stare at a person who is not looking at you, and notice that within several seconds that person will look straight back at you. Our reaction to a smile is also quite amazing. Most of us are very skilled at recognizing a forced smile, which utilizes a different set of muscles from naturally occurring smiles, but at the same time most people cannot explain why a forced smile looks different from a natural smile.

It became particularly clear how powerful a facial expression is in an interesting experiment recently conducted in Britain. The experimenters placed a coffee machine in a crowded public office and hung a sign above the coffee machine asking those pouring themselves a cup of coffee to insert a pound in a nearby box as payment. When the experimenters compared the number of cups that were poured from the machine with the amount of money in the box a week later, they discovered, unsurprisingly, that many people ignored the sign and made use of the machine without paying.

The second week, the experiments added a photograph of a pair of eyes looking directly at users of the coffee machine. This simple change had a large behavioral effect. A week later the amount of money in the box was very close to the number of cups of coffee poured from the machine.

To estimate the human capacity to identify the mental state of another individual, one researcher conducted an interesting experiment using the well-known British television game show called *Split or Steal.* The game involves a pair of players challenged to answer a series of trivia questions. For every correct answer the players receive a sum of money.

At the end of the round of questions, a decision has to be made regarding how to divide the accumulated money (which can sometimes be in excess of a hundred thousand pounds) between the two players. This involves each of the players secretly choosing one of two options, "split" or "steal." If both players choose split, they divide the accumulated sum equally. In contrast, if one player chooses split while the other chooses steal, the player who chose steal gets the entire sum of money while the other player goes home empty-handed. If both players choose steal, they both get nothing. Prior to choosing between split and steal, the players are instructed to conduct a thirty-second face-to-face conversation about what they intend to choose. This game is very similar to the famous Prisoner's Dilemma, which we will return to later in this book.

If you are a player on *Split or Steal,* from a purely monetary perspective it is always advantageous for you to choose steal. If the other player chooses split, then if you choose steal, you double your payoff.

On the other hand, if the other player is planning on choosing steal, then you won't get any payoff no matter what you choose, in which case you might prefer that your greedy erstwhile partner also go home without a penny. Absurdly, however, if both players fully understand this reasoning and act accordingly, neither of them will receive any money and will be forced to walk away from the hundred thousand pounds that they worked together to accumulate.

I highly recommend taking a look at video clips of the game show that are available on YouTube; they can be found by searching for the show's name, *Split or Steal*. During the short conversation between the players, they each make great efforts to persuade the other that they would never even conceive of choosing steal, since doing so would bring about the opprobrium of hundreds of thousands of television viewers, irreparably sullying his or her reputation. Many of the players on the show make this claim in a very convincing manner—only to reveal a few seconds later that they choose steal.

Einav Hart, a colleague of mine at the Center for the Study of Rationality, asked whether players could improve their ability to identify the mental states of others, thus increasing their chances of correctly predicting the choice of the other player. She showed video segments of the game to subjects who volunteered to participate in the experiment. Each subject was asked to predict the choice of one of the players, based on what was said during the conversation round, and was given a monetary reward for every correct prediction. After making their predictions, the subjects were further asked to predict the choice of the other player, but this time they were not incentivized with a reward. Hart showed that subjects were significantly better at making correct predictions when they received monetary incentives.

This result contrasted with several other research studies (mostly conducted by psychologists) that had concluded that people cannot reliably distinguish between authentic and faked emotional states. But if it were true that there is nothing that can be done to improve our ability to tell when emotional states are being faked, then monetary incentives should not have any effect. The fact that there was a significant difference in subject's capacity to make correct predictions as a result of giving them monetary rewards indicates that we do have hidden abilities to identify authentic emotions. These abilities apparently require a large effort of concentration and attention, which we are more willing to apply if we believe that a reward will follow the effort. It is possible that previous studies that failed to uncover our capacity to distinguish authentic from faked emotions did not sufficiently incentivize the subjects participating in the studies. In contrast to most experiments conducted by economists, including those run by the Center for the Study of Rationality, psychologists usually do not use monetary incentives in their experiments.

In the real world outside of laboratories, individuals receive a reward if they recognize inauthentic emotions correctly and receive a penalty if they are wrong (even if the reward is not necessarily monetary). This is why using incentives, as Hart did, is very important for studying human capacities for recognizing emotions. Hart's experiment tested the ability of observers of the game to recognize emotions; it is reasonable to suppose that players in the game should exhibit even sharper emotion-recognition capabilities.

Empirical research conducted several years ago in the United States by Avner Kalay also studied this topic.[3] Kalay studied data on the behavior of players in an American television game show called *Friend or Foe*, which is very similar to *Split or Steal*. In the course of his research he looked at hundreds of episodes of the game played over several years and noted the relative frequencies of the four possible outcomes in the game: (steal, split), (split, steal), (split, split) and (steal, steal).

Kalay discovered two striking phenomena. First, the frequency with which both players simultaneously made the same choice was high. In other words, (split, split) and (steal, steal) were chosen often and the other two possibilities were chosen infrequently. The second finding, which was even more striking, was that those players who chose steal earned—on average—almost the same as those who chose split. (I should emphasize that this is strictly an average. In any given game, the two players only get the same payoff if they make the same choice.)

This seems puzzling, since a few lines above we established that choosing steal gives you a higher payoff if the other player chooses split, while if he chooses steal it does not matter what you choose. How then is it possible for there to be no difference in the average payoff that the two choices provide the players in Kalay's study? The answer is simple. If you decide to play steal, you will not be able to conceal this decision fully from your partner. As a result, he will be more likely to choose steal as well, and you will be more likely to walk home with nothing. The net effect of all this is that, as Kalay's research showed, the two players tend to end up making the same decision, either split or steal. Even though both of them naturally claim during the conversation round that they have every intention to choose split, what they are really doing is making up their minds in the course of the conversation, while trying to read each other's thoughts. It is the human ability to judge emotions correctly that leads to the synchronization of the players' choices. As we have shown, this ability is very important in many situations, but as with many human abilities, not everyone is endowed with this talent in equal measure. Several years ago a well-known lawyer contacted me with a request that I serve as an expert witness on game theory for a company he was representing, which intended to establish a Web site for poker games. The law forbidding Internet gambling specifically prohibits games that involve chance as the main determinant of winning, but the law permits games in which skill is the main element of victory. If I could help persuade the court that poker is a game of skill rather than a game of chance, the legal prohibition on establishing a poker Web site would be lifted, and in return I would be paid a hefty fee.

I declined the offer, rather instinctively and decisively, but this decision might have been more emotional than rational. The truth is that poker is indeed a game of skill with a relatively small component of chance determining the winner of the game. The name of the game in poker is correctly judging the emotional state of your opponents.

When I was a child, my family, along with the families of my mother's seven siblings, would visit my maternal grandmother every holiday for a meal. After the meal the men in the family would all adjourn to the balcony to play a game of poker. We children would follow the game with great interest. At first we would conduct little wagers among ourselves over who would be the big winner and the big loser of the day. But we quickly realized that there was no point to wagering: Uncle Ezra was almost always the winner, while my father was losing almost as consistently.

Although the sums of money changing hands were insubstantial, the tension and emotions the game elicited were profound. Each round of the game ended with the players loudly expressing joy or anger. The game itself was played in absolute silence, with even the children holding their breaths.

My father barely glanced around him. He was entirely immersed in the cards, deciding which to play and which to discard, how much to raise the stakes, and when to fold. He nervously shifted in his seat while drumming his fingers on the table, waiting for the decisions of the other participants. Uncle Ezra, in contrast, was always calm and collected. He barely glanced at the cards in his hands, staring instead at my father as if he were curiously taking in his every move.

Sometimes my father tried to adopt Uncle Ezra's methods, but he never managed to mask his face with Uncle Ezra's coolly neutral expressions, nor did he ever succeed in reading Uncle Ezra's facial muscles to the point of being able to get a clue regarding Uncle Ezra's cards. Uncle Ezra was a much better poker player than my father was, due to his ability to identify the mental states of others while being able to hide his own.

The World Rock Paper Scissors Society is a similar case. It conducts a yearly international tournament drawing over five hundred participants who compete for cash prizes of up to $10,000. Most of us would think of the game rock paper scissors as a pure game of chance, but there are players who win at it consistently. The determining factor here is again the ability to recognize and hide intentions.

In social situations, which are much more complex than games of poker or rock paper scissors, the ability to identify the intentions of others requires much finer emotional insight and is strongly related to the ability to feel empathy for others.

Empathy, the capacity to experience the emotional experiences of others even when we are not acquainted with them (and even when they are fictional characters, as in a film or novel) is a wonderful phenomenon that has very ancient evolutionary roots. In 2004, an interesting research study was conducted in Italy, showing that monkeys tend to imitate the actions of other monkeys immediately after their birth, with no need for a lengthy learning process.[4] This ability was traced to "mirror cells" in the brain, which are responsible for imitation. Mirror cells are neurons containing electrical activity fired during the course of actions (especially motor actions). Interestingly, when another individual is seen undertaking the same actions, electrical activity is also triggered in the same mirror cells. For example, when a chimpanzee lifts his left arm, that action is brought about by electrical activity in neurons. In some of these neurons, identical electric activity is triggered when the chimpanzee sees another chimpanzee lifting his left arm, even if the watching chimpanzee is not lifting an arm and has no intention of doing so.

Scientists conduct more invasive research experiments on the brains of monkeys than can be conducted on humans. Electrodes implanted into the brains of monkeys can identify electrical activity down to the level of individual cells. The experimental evidence for the existence of mirror cells in humans is more indirect but still very persuasive. The evidence is mainly based on fMRI imaging indicating elevated oxygen consumption in various areas of the brain. The fMRI images show that the brain areas that exhibit activity when a person undertakes a particular motor action are also active when that person watches someone else undertaking that same action.

There is a broad consensus among brain scientists that empathy is a result of mirror cell activity. But in contrast to motor mirror cells, which are responsible for physical actions, empathy arises from emotional mirror cells. A research study conducted in 2009 using fMRI imagery showed that when children watch films depicting a person suffering pain, their brains exhibit activity in the same areas that would be activated if they themselves were feeling it. Research conducted on adults reveal similar phenomena involving brain activity when the subjects are shown pictures of people suffering sadness or fear.

The capacity for empathy is related to a very important concept used in philosophy and psychology called Theory of Mind (ToM). ToM is not a scientific theory; it refers to the human ability to have beliefs regarding the emotional states, beliefs, and intentions of another person. ToM is regarded as a very important characteristic distinguishing humans from other living creatures. It is identifiable in children as young as two years old, who turn their gazes toward the same items in a room that other people around them are looking at.

The capacity for ToM improves considerably at age three or four; a child at that age can often distinguish between what he or she knows versus what others know. If you wish, you can try the following experiment on a four year old. Take two boxes colored in different colors—for example, one red and one yellow—along with a small candy bar. In the presence of the child and another adult, place the candy bar in the red box and ask the adult to step out of the room. After the adult has stepped out, but in the presence of the child, move the candy bar from the red box to the yellow box. Now invite the adult back in and ask the child in which box the adult thinks the candy bar is located. If the child gives the right answer and points to the red box, then he or she is exhibiting healthy ToM. Children with autism spectrum disorder, who have inhibited ToM, may not pass this test even at a much older age.

I intend soon to begin conducting a research project along with several psychiatric researchers in an effort to improve our understanding of ToM with the help of game theory. Games such as the ultimatum game and the trust game (which will be discussed in later chapters), as well as other games, may enable us to identify children who have slight deficiencies in attaining ToM or who have mild autism spectrum disorder but otherwise successfully pass standard tests. The capacities for empathy and ToM are related because both of them are based on the ability to put oneself in another's shoes.

If you shut your eyes for a moment and try to imagine yourself with the same cognitive capacities that you now have but with no ToM, you will discover that this is a very frightening image—indeed, lacking empathy and ToM are common symptoms of autism spectrum disorder, which account for many of the daily difficulties faced by people with this disorder. In such a state, although you would be fully conscious and aware of your surroundings, you would in a sense be living on another planet populated by alien life-forms whose behaviors and reactions would be completely unpredictable. You would have no way of knowing whether or not scratching your left ear would deeply insult them and elicit an aggressive response. You would not know how to gain their trust or persuade them to help you find food. Even if they approach you with benign intentions, you would not know whether they come in peace or with the objective of attacking you, and you would certainly be unable to form an intimate bond with any of them or have offspring with them.

Just as you can't identify which muscles are involved in a person's fake smile, it is all but impossible to use logic to discern another person's emotional state, and thus their intentions, whether in a game of poker, at the negotiating table, or on a date. The ability to read people and send back the signals they want is irreducibly emotional, and at the same time is an essential piece of data in making a good decision: whether to fold or raise, offer a compromise or hold firm, expect a kiss or a smack in the face. Our emotions contain arguments and facts we simply can't relate to in any other way.

4

GAME THEORY, EMOTIONS, AND THE GOLDEN RULE OF ETHICS

THE PRISONER'S DILEMMA IS PERHAPS THE MOST OVERUSED PARADOX in all of the social science literature, but it continues to fascinate anyone, whether a professional researcher or not, who finds himself caught within its slippery traps. Can introducing rational emotions into the equation help us to find our way out of the paradox?

Let's briefly review the elements of the Prisoner's Dilemma. Two suspected bank burglars are caught and arrested. The police, however, lack sufficient evidence. Without a confession from at least one of the suspects, the police will have no choice but to release them.

Each prisoner is held in isolation in a separate cell. The police interrogator calls each one in turn into the interrogation room and offers him the following deal: if one of the two of you confesses while the other refuses to confess, the confessing one will be released. The refusing one will be convicted and serve a five-year sentence in prison. If both of you confess, you will both be convicted, but in return for your confessions we will go easy on you, and you will be sentenced to only four years. The prisoners also know that if neither of them confesses, the police will be unable to charge them with burglary but only with reckless driving during the chase, which will lead to a one-month prison sentence.

Each of the two prisoners must decide how to respond to this offer—but without having any opportunity to coordinate in any way with the other prisoner locked away in the separate cell. Will the prisoners confess or not?

Place yourself in the shoes of one of the prisoners, and you will quickly realize that it is always in your best interest to confess, regardless of what you expect your partner in crime will do. If she confesses, then your confession along with hers reduces your prison term by one year (from five years to four years). If she refuses to confess, your confession buys you an immediate release to freedom.

This result, however, is paradoxical. The two prisoners both find themselves using rational and selfish considerations to conclude that they should confess to the crime, leading to four-year prison terms for each of them. But if they were both to refuse to confess instead, they would both be much better off, serving only minor one-month sentences.

The Prisoner's Dilemma is not an idle intellectual amusement—it is a core concept in game theory. Game theory is, in essence, the study of interactive decisions. A "game," in professional jargon, is any situation in which the actions of one person affect the situation of another person. Economic competition, violent conflict between nations, and even interactions within families can all be modeled using game theory.

The Prisoner's Dilemma is often called the social dilemma game by social science researchers, because it succinctly describes a wide range of social and economic situations, including environmental pollution, tax evasion, military draft dodging, and even cutting in line at the bank. In all of these cases there is an action that is preferable from the perspective of each isolated individual. However, if all (or even a majority) of the participants do undertake that action, then everyone ends up suffering. In the real world, how should we resolve such dilemmas? What makes people prefer to cooperate in such situations even when there is no way to impose cooperation?

One answer to this question was given by Robert Aumann in a series of research papers for which he was awarded the Nobel Prize in economics in 2005 (the prize was co-awarded to Aumann and Schelling that year).[1] Social variations on the Prisoner's Dilemma are frequently "repeated games," in which the same interactive situation is repeated a large number of times involving the same group of players. The repetition makes choosing a selfish action potentially very costly: people remember your behavior from past occasions. Thus, a player who acts selfishly (such as by confessing in the classic Prisoner's Dilemma) is liable, when the same situation arises again, to be punished by the other players, who will then also choose actions that are favorable to their purely selfish interests (such as confessing themselves).

Aumann constructed a mathematical model of repeated games and showed that in repeated situations it is possible to achieve cooperation through rational considerations. Aumann's theory deserves a full discussion on its own, and it will be presented in greater detail in the next chapter. There is, however, another possible answer based on my own research.[2] Understanding it requires introducing one of the central concepts of game theory, "Nash equilibrium," named after John Nash, who won the Nobel Prize in economics in 1994 and became world famous as the central figure in the film *A Beautiful Mind*. Nash first had the idea for his equilibrium concept in the early 1950s; it eventually became an extremely important concept that is used throughout the social sciences.

To explain the idea behind Nash equilibrium, we will concentrate on games involving two players. Each player has a list of actions (or strategies) that he or she may make use of. A pair of such actions, one chosen by each player from the list of available actions, determines a payoff for each player. An equilibrium is achieved if the action chosen by each player is the "best reply" to the action chosen by the other player. In other words, neither player can improve on the payoff received by choosing a different action.

To take a more specific example, consider a game called the battle between the sexes: you and your spouse need to decide where to go out tonight.

There are two possibilities, a ballet production or a boxing match. Unfortunately, you and your spouse have divergent preferences: you insist on an evening at the ballet, while your spouse refuses to give up the opportunity to enjoy a good boxing match.

After a lot of fruitless deliberation you decide that the choice will be determined in the following manner. Each of you will write down either "ballet" or "boxing" on a slip of paper, without knowing what the other wrote down or discussing the matter between you. The slips of paper will be handed to your neighbor, Mrs. Brown, promptly at 7 p.m. Mrs. Brown will then read aloud what was written on the slips. If both of you wrote down the same activity for the evening, then that is what both of you will do together. On the other hand if you wrote different activities, you will both stay home, missing out on a night out. Suppose now that you each rate your preferred activity as worth $200, while you consider your less-preferred activity as worth $100. Staying at home is the worst option from your perspective, worth $0. What is an equilibrium in this game?

The only possible equilibrium here is achieved if either both of you write down "ballet" or both of you write down "boxing"; if you both insist on writing down your most preferred activity, then you will end up staying at home. It follows that the only way to improve the situation is if one of you gives in and agrees to go to his or her less-preferred activity. But this is where the "trap" in this example lies: if both you and your spouse decide to be magnanimous and give in to the other's preferences, you will end up staying at home (recall that you are not permitted to discuss with each other what you are going to write down).

Can our couple improve on their chances of arriving at an identical choice of activity that will guarantee an evening out? Of course! For example, the boxing fan might place a boxing glove on the dining room table as a sharp hint that she has no intention at all of backing down from her preferred activity, whatever the consequences may be. That might well have the effect of persuading the ballet aficionado that unless he wants to spend the evening at home, he has no choice but to go along with his wife's preferences, increasing the probability that he will write down "boxing match" as his choice.

Alternatively, the husband might wish to preempt such a move by his wife by loudly playing the music from Tchaikovsky's *Swan Lake* in the living room to signal that he is sticking to ballet as his choice, come what may, thus increasing the chances that his wife will buckle and match his choice by writing down "ballet" as her choice.

Lacking the option of speaking to each other directly, the couple may indeed resort to such signals as a way of improving the chances of arriving at an equilibrium in the battle of the sexes game. But how is this related to emotions?

Emotions are, in effect, a signaling mechanism that enables us to coordinate our actions and arrive at an equilibrium in a wide range of games in which we participate on a daily basis. Emotions also enable us to create new equilibria that could not exist in a world of pure thinking and reason. In many cases they improve our social situations through this mechanism.

To understand this important point, let's go back to the Prisoner's Dilemma and show how emotions can create a cooperative equilibrium even when the game is played only once. For this purpose we will describe the Prisoner's Dilemma in a slightly different way:

Imagine that you and a complete stranger participate in an experiment. You are each initially given $100. You are then asked to choose one of two possible actions, "take" or "be generous" (the two of you have no opportunity to discuss your choices prior to making them). If one of you chooses take while the other chooses be generous, then the one who chooses be generous is required to transfer the entire sum of $100 to the taker. If both of you choose take, then each of you will be required to return $50 to the experimenter. Finally, if both of you choose be generous, you will each receive an extra $50 from the experimenter, and both of you will go home $150 richer.

Note the resemblance of this game to the game *Split or Steal* mentioned in the previous chapter. Here, as in that game, if all you care about is attaining the highest possible monetary reward, then you should always choose take. That choice will always give you more money, regardless of what the other participant chooses.

Now let's add emotions to this game. Suppose that in addition to the monetary payoff that you receive in the game, you also place a value on being a decent and fair individual on the one hand and on not being "a pushover" on the other. If you choose take while the other player chooses be generous, then you will feel ashamed at your greediness; that sense of shame has some negative value, which we choose to describe here as equivalent to losing $100. On the other hand if you choose be generous while the other player chooses take, then you will feel a sense of insult and anger, which we can also say is equivalent to being penalized $100. If both of you choose take or both of you choose be generous, then you will experience a neutral emotional reaction.

Now, assuming that the other player has identical emotional reactions in these situations, with identical monetary values attached to them, the analysis of the game changes significantly. Whereas the best case of the take option was once $200 in cash, now it is $100, because of what you might call the shame penalty. This new value is lower than what you get for be generous, meaning $150 is your best-case scenario. Thus the two players' simultaneous choice of be generous becomes a new equilibrium, meaning both parties are likeliest to cooperate as opposed to being selfish.

Put very simply, this means that the presence of emotions in the equation, even negative ones like anger and shame, can lead to better outcomes for both players. This explanation, however, is still incomplete. I want to show that the emotions described in the above example were not arbitrarily chosen and that in fact they serve the narrowest material interests of those who feel those emotions.

Suppose that having emotional responses also gives a player a fairly good ability to anticipate the emotional reactions of others. Next, imagine what would happen if one of the players in the Prisoner's Dilemma game is an emotionless individual who employs only the coldest of considerations aimed at maximizing the amount of money he gets, while the other player has the reasonable emotional responses (along with the emotional anticipation capabilities) described above. Let's suggestively call the cold and calculating player Mr. Brain while the other player is Mr. Emotions.

Mr. Brain will certainly choose take, since he has no shame. But Mr. Emotions is likely to recognize that he is facing off against Mr. Brain and therefore to anticipate that Mr. Brain will choose take. In such a situation, if Mr. Emotions chooses be generous, he will lose twice: once when he loses the $100 given to him at the start of the game, and once when he feels insulted, which is equivalent to a further penalty of $100, leading to a total loss of $200. If, on the other hand, he chooses take, he loses only $50. Mr. Emotions thus concludes that he should also choose take, and both he and Mr. Brain walk away with $50. This contrasts with the situation that holds if both players are emotional players, which as we showed above enables an equilibrium that grants them $150. The conclusion is that emotional behavior is advantageous: in this simple example there is a positive monetary advantage to emotional responses.

This example is taken from a mathematical model I have developed that generalizes the concept of Nash equilibrium. The model shows that the main motivation for cooperation in many games similar to the Prisoner's Dilemma is an emotional need for reciprocity, such as a feeling of shame at an expression of greediness when others are being generous or a sense of anger and insult in the face of greediness on the part of others. This pair of emotions come together to form the golden rule of ethics, which is also called the ethic of reciprocity.

The golden rule is much touted in our religious texts and taught to every schoolchild as a means to protect the emotions of others— something you have to do even though it runs against your personal desires. But as these experiments show, it is just as importantly a means to look out for our own narrow interests.

5

THE PRISONER'S DILEMMA IN REPEATED INTERACTIONS

Do Drawn Knives Increase Cooperation in the World?

SPONTANEITY, AUTOMATIC RESPONSE, AND SPEED OF REACTION ARE among the most important characteristics of emotional responses. In fact, there are many cases in which sheer speed of reaction is one of the advantages that emotional responses have over careful deliberation. Our instinctive recoil on seeing a snake crawling in the grass saves us from possible danger far more efficiently than would cognitive analysis of the situation.

The speed and automatic nature of our social reactions are, it turns out, very important. I will show in this chapter how emotional behavior, perhaps paradoxically because of its automatic nature, can bring about cooperation in situations in which rational behavior fails to do so.

We will take another look at the Prisoner's Dilemma, but this time we will concentrate on situations in which the players play the same game many times. This means that the players need to take into account long-term strategic considerations.

In the previous chapter we showed that rational and selfish individuals will not cooperate in the Prisoner's Dilemma game when it is played only once, since noncooperation is a so-called "dominant strategy"—it guarantees a higher payoff no matter what the other player does. Now consider what happens if the game is played twice. At each stage of play, each player decides whether to cooperate ("be generous") or not to cooperate ("take") at that stage. After both stages are completed, the total payoff that the players receive is the sum of the payoffs in each of the two stages.

To analyze rational behavior in this repeated game, we start by concentrating on the second stage of the game. In the second stage, the original Prisoner's Dilemma is in effect being played only once—there is no next stage in which to punish or reward behavior this time around. Thus the strategic analysis is equal to that of the one-stage Prisoner's Dilemma, which we have already shown leads to the conclusion that the only rational behavior is noncooperation by both players.

Knowing what rational players will do in the second stage, we can try to predict how the players will behave in the first stage of the game. The players' behavior in the first stage has no influence on their payoffs in the second stage of the game, hence the first stage is in effect a one-stage Prisoner's Dilemma as well. In the first stage the players will once again both choose not to cooperate.

It is not difficult to see that the same reasoning will apply to any number of repeated stages, as long as both players know the exact number of stages that will be played, whether it is one or three or a hundred thousand. In greater detail, when both players know that they are playing in the last stage of the game, they have no rational reason to cooperate no matter what happened in the previous stages. But it then follows that in the second-to-last stage they will not cooperate, and so on. This sort of reasoning is called an inductive argument, and it is often used in game theory.

Notice that the induction here starts off with both players not cooperating in the last stage. But what happens if the players *don't know* when the last stage will be, even as it is happening? Most human in-teractions are, in fact, like this. Consider for example the interactions you have with your regular car mechanic, your colleagues from work, or even your spouse. You almost never know exactly how many more times in the future you will interact with them. This naturally leads to the question: What rational behavior can be expected without the assumption that the players know when they will reach the last stage of a repeated game?

Robert Aumann answered this very important question. It is con-sidered one of his most important contributions to game theory. Us-ing a mathematical model, Aumann proved that in such situations cooperation is possible in equilibrium even when the players are ra-tional. Both the model and Aumann's proof are beautiful and deep constructions. Fully explaining them in detail requires delving into a level of formal mathematics that is beyond the scope of this book, so let me try to describe them in plainer terms.

Imagine yourself playing the Prisoner's Dilemma repeatedly in a situation where after every stage there is a 99 percent chance that you will play the same game again against the same player and a 1 percent chance that you will never meet that same person again. This descrip-tion is a bit unrealistic—it probably overstates the number of interac-tions you're likely to have in the long run with any one person. But it is valuable for describing the short-term mindset of most interactions, so let's put that objection aside for now.

We need to consider what "strategy" means in this case. In the one-stage game a strategy is simply a decision of whether or not to cooperate. In a repeated game the concept of a strategy is much more complicated; it is in effect a thick book of decisions, with each deci-sion relating to the action that you will choose given what has hap-pened up to now in the play of the game. Here is an example of such a strategy: up to the 700th stage I cooperate no matter what the other player has done, and from the 700th stage onwards, after every stage that the other player does not cooperate, I reciprocate by not cooper-ating for the next two stages.

If you think that this seems like a complicated strategy, my response is that it is actually a very simple strategy—note that I managed to describe it entirely within a sentence and a half. There are strategies that are so complicated that to write them down even for the first few stages I would need more paper than can be found in the entire Library of Congress (including the paper in the restrooms). Often, however, the most complicated strategies are the least interesting. In fact I will describe in this chapter two strategies that are extremely simple but of great interest. They are:

1.   The Grim Trigger strategy—in the first stage I choose "be generous" and continue to be generous as long as the other player also chooses "be generous." However, if the other player chooses "take" at some stage (even if the other player does this only once) then I will forever after that choose "take" in every subsequent stage.
2.   The Tit-for-Tat strategy—at every stage I choose the same choice that the other player chose in the previous stage.

Two rational players (whose only goal is their personal material gain) who both use the Grim Trigger strategy will find themselves in an equilibrium under which they will both cooperate (choose "be generous") forever. The explanation for this is quite simple. First note that if both players are using the Grim Trigger strategy, then they cooperate in the first stage. Both players will recognize that the other player has cooperated. The strategy will then lead both of them to cooperate in the second stage, and similarly to cooperate in the third stage, and so on. At each stage in which they both cooperate they each add $50 to their total winnings.

Neither player can do better than this by choosing a different strategy, as long as the other player sticks to the Grim Trigger strategy. It is true that if one of the players were to choose "take" at some stage, when the other player is using the Grim Trigger strategy, then the player who chose "take" will get $100 at that stage, making him better off by $50 than he would have been had he chosen "be generous." But by doing so he will have triggered the other player's "punishment": in every subsequent stage (and there are many subsequent stages to be expected) he will lose $50 instead of gaining $50, as the other player steadfastly chooses "take" no matter what happens. Note that stable cooperation is attained here due to the fact that any noncooperation triggers immediate noncooperative retribution on the part of the other player, creating a situation of effective deterrence to noncooperation.

✳ ✳ ✳

IN HIS NOBEL PRIZE ACCEPTANCE LECTURE IN STOCKHOLM, SWEDEN, Robert Aumann spoke about a game theoretical insight very similar to the one presented in the previous chapters. He even claimed that this insight explains the essence of nearly every international conflict, including the Israeli-Palestinian conflict. The message was that to prevent bloodshed, people need to create mechanisms for deterrence using tough strategies, as the United States and the Soviet Union did during the Cold War. Only strong deterrence, under this argument, can prevent people from succumbing to incentives for conflict.

Shortly after Aumann's Nobel Prize ceremony several media publicists approached me and asked me to respond to this claim. I argued that despite the fact that the insight presented by Aumann is deep and beautiful and that I couldn't point to anyone who deserved a Nobel Prize more than Aumann, there are very few direct connections between the elegant mathematical results in this field and concrete applicable conclusions for international conflicts. Deterrence alone is a situation too unstable to be used as a dependable basis for maintaining peace and preventing bloodshed—any small change may set off the "Grim Trigger." Although the theoretical model implies that cooperation is an equilibrium under conditions of deterrence, once that equilibrium is broken the entire edifice on which peace and cooperation depends is shattered because the very threats underpinning the deterrence are liable to lead to disasters on global scales (imagine what would have happened if the United States and the Soviet Union had actually implemented the bellicose threats that they regularly issued during the Cold War).

Deterrence alone is not sufficient. Alongside deterrence based on threats we need to construct systems that include positive inducements for both sides, such as joint economic interests, for example, that can serve as an additional source of stability in international relations. This is similar to the idea that individuals should be motivated using both carrots and sticks.

Some people went much further than I did in opposing some of the ideas presented in Aumann's Nobel lecture. A group of Israeli leftists formally petitioned the Nobel Committee with a request that it withdraw the awarding of the Nobel Prize to Aumann due to his political opinions and the political lessons he draws from his scientific research. This infuriated me (possibly an irrational emotional response). If science were administered along strictly politically correct lines and its leading practitioners were only rewarded based on their political opinions, human progress would still be stuck in the same place that it was during the Dark Ages.

The Tit-for-Tat strategy is less drastic than the Grim Trigger strategy but still ensures equilibrium. The Tit-for-Tat strategy also punishes noncooperation on the part of one player, but in this case the punishment for noncooperation that lasts only one stage is more forgiving than the punishment of the Grim Trigger strategy. If the noncooperative player goes back to cooperating in the next stage, then the punishment ceases and the players go back to playing cooperatively in each stage.

It turns out that Tit-for-Tat leads to a cooperative equilibrium; neither player can profit by unilaterally choosing not to cooperate. If a player does choose not to cooperate for a few stages and then cooperates again, the play of the game will go back to a cooperative path in the future, but until that happens she will lose more than she gained from temporarily not cooperating. (It takes a little math to show this, but see for yourself if you like. What happens if a player chooses not to cooperate for only one stage? How much does she gain that first round, and how much does she lose thereafter?)

We have so far considered repeated interactions in which, after each stage, both players expect that there will be another stage of the game with high probability. What happens in other situations? Consider two concrete examples. Imagine yourself enjoying a week of vacation in Malaga, Spain. On the first day of the vacation you walk into a restaurant and are so pleased with the excellent meal served there that you decide to return and eat there on every remaining day of the vacation. Each time you sit at the restaurant the same waiter serves you. In this scenario, your interactions with the waiter are in effect in a six-stage (six being the number of remaining days in the vacation) repeated Prisoner's Dilemma.

Cooperation, which involves the waiter giving you good service and you reciprocating with a generous tip, is significant in this situation. Note that on each day of your vacation—except for the last one—you expect repeated interaction with the waiter with high probability. On the last day, however, you expect with high probability that you will not be returning to the same restaurant at any time in the foreseeable future, since it is the last day of the vacation, your flight tickets were booked long ago, and you need to be back at work the day after tomorrow.

Can the Grim Trigger strategy ensure a cooperative equilibrium on each day of the vacation? Clearly not (again, assuming rational considerations with a single goal of selfishly maximizing your material condition). Even if the waiter is under the impression that you will be staying in the city for a very long period of time with an uncertain last day, cooperation will not be maintained on each day of your vacation for the simple reason that on the last day of your vacation you have no (selfish) reason for leaving the waiter a tip. There is a very small probability that you will return to the same restaurant the next day (your flight might be canceled, so we can suppose that probability is small but not zero). It follows that if you walk out without a leaving a tip, the probability that the waiter will be able to punish you with bad service in the future is very small.

If the waiter is sufficiently rational, intelligent, and "selfishly materialistic," he will understand that at some point there will come a day in which you will walk out of the restaurant without leaving a tip even if he gives you first-class service. That might be enough to wipe out the incentive he has to serve you well every day: he knows with certainty that a day without a tip is coming, he just does not know exactly when that day will arrive.

This description of the peculiar relationship between a vacationer in Malaga and a local waiter might seem a bit over the top, but it actually plays out this way more often than you might think. It is known that people tend to give larger tips in local restaurants where they are regular customers than in foreign restaurants that they happen to stumble upon and that they will return to with very small probability. Service is also usually better in restaurants whose customers are local residents who frequently eat there, compared to tourist traps.

Despite this, we still often leave tips, even in places where there is no material gain for us in doing so. Why do we do this? Why do we avoid the opportunity to exploit the "last day effect" cynically whenever we can? (In fact, there are people who tend to leave an especially large tip on the last day of a vacation as a form of gratitude for the good service they received over several days.)

The answer, unsurprisingly, lies in our emotions. Remember, in the real world we play out our Prisoner's Dilemma–like situations over and over again, not just once. To help think about this, let me introduce the concept of automatons.

Computer scientists invented automatons, but they are widely used in many models in economics and game theory. My small contribution to their work is this: I believe that emotions can be described using automatons and that this can lead to new insights, even though automatons are machines.

Automatons are defined using the following components (and only these):

1. A set of states
2. A set of actions
3. An outcome function that, given a state and an action, de-
    termines a new resulting state
4. An action function that associates each state with an action
5. An initial state

A copy machine that makes one hundred copies is a good example of an automaton.

Its set of states is the set of all the integer numbers from zero to one hundred (hence there are 101 states).

Its set of actions contains only two actions, "copy" and "stop."

Its outcome function takes every state x (between zero and one hundred) and returns the state $x+1$ if the action is "copy." If the action is "stop," the function returns the state x, that is, the state does not change.

Its action function returns "copy" for every state that is less than one hundred and returns "stop" when the state is one hundred.

Its initial state is zero.

You can see that based on the way it is defined, this automaton will start at state 0, then move on to state 1 followed by state 2 and so on. At each of these states the automaton will make a copy of the document until it reaches the state 100, at which point it stops. (If this description reminds you of a computer program, there is a good reason for that. An automaton is essentially a simple computer program.)

You might think of automatons (and computers) as the exact opposite of emotional beings, yet they are similar in at least one way: if you know the circumstances, they are predictable. If I react emotionally to the situations in which I find myself and always draw a knife when insulted, then my behavior can be described using only two states: (1) I feel insulted, and (2) I do not feel insulted. My action function causes me to draw my knife if (and only if) I feel insulted. I am, in effect, an automaton, and not even a very complex one.

In contrast, if I am a purely rational person, then my behavior will be more complex. A sense of insult alone might not suffice to cause me to draw my knife. I might do that only if I feel insulted and I am also persuaded that the person who insulted me cannot later prove in a court of law that I used a knife against him. The sub-situation in which it cannot be proved that I used a knife is itself composed of many other sub-situations (who else is in the area and can serve as a witness, is there a surveillance camera in operation that can be used in court, etc.). We see that the number of states needed to describe the behavior of a rational person is far greater than the number of states in the description of an emotional person, making the use of automatons for modeling rational behavior much more difficult. (Remember, emotions are adept at creating commitment—we are less likely to respond to such subtleties as the presence or absence of criminal witnesses when feeling insulted or angry.)

Hence, the crucial difference between a rational and an emotional reaction is that the latter is less dependent on the circumstances. This doesn't mean that an emotional person will always react in the same manner to an insult, but it does say that a rational person's reaction will be more dependent on the circumstances of the event (this is also consistent with the fact that a rational state of mind is associated with more self-control).

The emotional, "automaton" description feels a bit more like real life, doesn't it? You might be puzzled regarding the "drawn knife" example above—after all, drawing knives could not possibly lead to useful cooperation. But that is wrong. The emotional behavior that leads to drawing knives is a positive element in forming cooperation. To be more precise about this, and to avoid exaggeration, let's state it this way: vengeful behavior, in the right dosage, can be a positive element in forming cooperation. Hesitant and over-forgiving emotional behavior will not lead to cooperation. To the contrary, it leads to egotism, because in a world in which every action is forgiven, every individual has an incentive to act egotistically while harming others.

Imagine that you are the following automaton playing the game:

1. The set of states represents your emotional state: you are either angry or calm.
2. The actions are either "cooperate" or "don't cooperate."
3. The outcome function takes the action chosen by the other player in the previous stage and determines your state in the current stage as follows: if the other player chose "cooperate," you are now calm, but if the other player chose "don't cooperate," you are now angry.
4. The action function takes your state into account and determines the action you choose as follows: if you are calm, then you choose "cooperate," and if you are angry, you choose "don't cooperate."
5. Your initial state is "calm."

If both players are automatons as described above, then they will definitely cooperate throughout all the stages of the game. This follows because they both start in a calm state, leading each of them to cooperate, which further keeps them both in a calm state and so on; no player will ever be angry.

We need to check whether a player can gain more by behaving as if she were a different automaton (assuming that she is playing against the automaton described above). For example, we can imagine that one of the players is always in an angry state no matter what happens or is always in a calm state no matter what.

To gain more, even in the short term, a "deviating" player will need to choose "don't cooperate" in at least one stage, giving him a payoff of $200 as opposed to $150 (since his opponent will have chosen "cooperate"). But this behavior will have implications in the later stages of the game. After the deviating player has chosen "don't cooperate," the other player will be in an angry state, leading her to choose "don't cooperate" in the next stage. If the deviating player chooses "cooperate" at that stage then he will get $0 instead of $150, so that he ends up losing more than he gained from his one-time deviation. If the deviating player instead continues to choose "don't cooperate," in subsequent stages he will lose $100 (relative to what he could gain if he always chose "cooperate") every time he does so.

The only opportunity for a deviating player to gain a profit is when his behavior has no implications for the future—which is when there is no relevant future, that is, in the very last stage of the game. But if the deviating player is a two-state automaton whose state depends only on the actions of the other player (meaning that he exhibits emotional behavior), his actions cannot depend on which stage the game is currently in. We conclude that an emotional player cannot improve his total payoff by acting differently from the automaton described above. It follows that cooperation at each and every stage forms an equilibrium.

The interesting point here is that each of the two emotional players in this situation will earn more under equilibrium than either would if both players were rational players playing the same game. From this perspective, emotional behavior is better for sustaining cooperation in the repeated Prisoner's Dilemma game even when the number of stages of the game is definitely known by both players.

Now, let's return to our Spanish waiter and why you tip him. In your interaction with the waiter each of you behaves like an automaton with two possible actions: "tip" and "don't tip" for you and "provide good service" and "provide bad service" for the waiter. Every day each of you is controlled by one of the following emotional states: "anger" and "happiness." These states are determined by the recent action of the other party. You are happy if you got good service, and the waiter is happy if he gets his tip. Finally, a state of happiness drives you to leave a tip and drives the waiter to provide good service. All this dictates a dynamic in which the date (i.e., whether it is the last day of your vacation) plays no role. You and your waiter are simply automatons too simple to get the date into the equation. If you are an emotional automaton, as so many of us seem to be, then you will be rewarding him for *today's* service, and he will reward you with quality service in response to the tip you left during your last visit to the restaurant. The fact that today is your last day in Spain doesn't matter; you will only punish bad service.

If you thought of feeling insulted by this description, you shouldn't. You are conscious and smart enough to know the date and whether it is the last day of a vacation in Spain, but your emotional state prevents you from making the link between this piece of information and the decision of whether or not to leave a tip.

What would happen if one of you, say you yourself, is perfectly rational (and selfish) while the other is an emotional automaton like the one described above? You would still tip your waiter on each day except for the last one. Failing to do so will trigger bad service tomorrow, but you won't be there. But if both of you are perfectly rational, the waiter can expect you not to give a tip on your last day, and therefore will offer bad service. As argued earlier in the context of the Prisoners' Dilemma, your cooperation is doomed to fail. You will leave no tip, and you will get lousy service throughout your entire vacation.

The main take-home insight from the entire analysis here is quite surprising: it is simplicity and straightforwardness, rather than sophistication and subtlety, that are conducive to cooperation and eventually to making both parties in an interaction better off.

6

ON DECENCY, INSULT, AND REVENGE

Why Don't Suckers Suffer from Disgust?

THE 1994 NOBEL PRIZE IN ECONOMICS WAS AWARDED TO REINHARD Selten, along with John Nash, for his contributions to game theory. Selten developed a dynamic equilibrium concept in which players think forward in a manner similar to the way chess and checkers players try to think several moves ahead.

Selten's student Werner Güth conducted a simple experiment in 1982 called the ultimatum game.[1] In this game two players divide a sum of money, say $100, between them based on the following rule: the first player offers the second player a sum of money from the $100 (he can offer to give anything from $0 to the entire $100). If the second player accepts the offer, then the $100 is divided among the players according to the terms of the offer. If the offer is rejected, the experimenter takes away the $100 and both players walk away with nothing. The first player's offer is in effect a "take it or leave it" ultimatum, explaining the game's name.

Two selfish and rational players playing this game will agree to a division granting the proposer $99 and the responder only $1. Since the game is played only once, the responding player should accept whatever nonzero sum of money is offered to him since even one dollar is preferable to getting nothing at all. The proposer, knowing this, should put forward the lowest possible offer, one dollar.

That is what Selten's model of equilibrium predicts would happen. But Selten (whom I had the privilege of working with for two years), is not only a great scientist, he is also a man of outstanding intellectual integrity. He felt dissatisfied with the equilibrium concept that had earned him an international reputation and eventually won him the Nobel Prize. Selten predicted that actual plays of the ultimatum game will usually result in a division of the money completely different from his equilibrium.

Güth's experiment, conducted in Germany with a large number of participants, revealed that in most cases the money was divided 50–50 among the two players. In addition, most offers made by the first player that amounted to 35 percent or less of the money were rejected by the responding player. In other words, the responding player was usually willing to give up the opportunity to receive $35 as long as this resulted in the offering player not receiving the $65 that he had greedily wanted to take for himself.

Hundreds of articles have been written about the ultimatum game since Güth's famous results were published. Researchers in economics, business administration, political science, psychology, anthropology, and philosophy have written on the subject. Many research studies have compared how players in different cultures behave in the ultimatum game, including African tribes and isolated tribes in the Amazon River basin. A group of researchers at the Max Planck Institute in Germany even published an article in 2007 on the subject of how chimpanzees play the ultimatum game.[2] (In case this sounds improbable, here's how it worked: The chimpanzees were sitting in separate cages facing a device with two pairs of trays: one with five bananas to each chimpanzee and the other with nine bananas to chimpanzee A and only one to chimpanzee B. The device allowed chimpanzee A to pull whichever pair of trays it chose, but it could only pull these trays halfway. For the bananas to be claimed, chimpanzee B had to agree to this choice and do its own pulling.)

The ultimatum game has attracted attention in fields far removed from pure game theory because it deals with a question that is elementary and very important for all the social sciences: How applicable is the assumption that individuals are selfish and rational, keeping in mind that this assumption underpins most theoretical models in economics and many of the social sciences?

Variations of the ultimatum game have been studied to gain insights into the differences in the reasoning used by the proposer versus the reasoning of the responding player. An offer of a 50–50 split of the money on the part of the proposer might be motivated either by a desire for equality and decency or by fear that the responder will spurn an offer that is too low. To ascertain the true motivation of the proposer, researchers have suggested studying behavior in the dictator game, as mentioned in the previous chapter, instead of the ultimatum game. In the dictator game the second player must accept the first player's offer; she has no recourse to take revenge against an insultingly low offer by denying both himself and the first player any monetary reward.

If players who offer a 50–50 division in the ultimatum game also make the same offer in the dictator game, then we can deduce that their primary motivation is a desire for equality, because in the dictator game the second player has no recourse to punish the first player. If, on the other hand, they switch to giving low offers in the dictator game, then that would be a strong indication that their primary motivation for offering a 50–50 split in the ultimatum game is a fear of losing all the money due to the possible revenge of the second player in response to a lower offer, rather than any desire for equality. The results of experiments in which players play both the ultimatum game and the dictator game show that player behavior in the ultimatum game is quite rational: players learn how to predict the reactions of the other players and seek the lowest offers they can get away with, without triggering rejection on the part of the other players, in order to maximize their profits.

Many important insights have been revealed by comparing the behaviors of people from different cultures when they play the ultimatum game. One published research paper on the subject compared ultimatum game players in the United States, Japan, Slovenia, and Israel.[3] The research study found significant differences between different cultures, whether players were in the role of proposers or responders. Players in Israel tended to propose the lowest offers for dividing the money. Japan was not far behind Israel, in second place in terms of the selfishness of the offers made by proposing players. Players in Slovenia and the United States were much more generous in their offers.

The most astonishing result of this intercultural comparative research study, however, was the close correlation between the offers made and the responses to them. In both Israel and Japan responders tended to accept relatively low offers. But when similar offers—and even more generous offers—were made by proposers in the United States, they were often summarily rejected by responders.

The conclusion we take from this experiment is that norms of what constitutes fairness are relative and culturally determined. An offer considered fair in Japan or Israel may be construed as outrageously low in the United States. Conversely, a normal offer in the United States may be seen as generous (or even a "sucker's offer") in Israel. An offer deemed unfair by both cultures will almost always be rejected. Even Israelis, the least prideful group in accepting money in this game, tend to reject offers of 20 percent or less, but their threshold of acceptance is lower than that of Americans.

Proposing players "magically" know what constitutes fairness in their own cultures and try to make the lowest offers that will likely be accepted by responders. Their behavior is very consistent with assumptions of selfishness and rationality. This ability to read signals of fairness, as we saw in Chapter 5, is one of the important virtues of rational emotions. It eliminates lots of unnecessary disagreement and wasted time.

Several years ago my colleague Shmuel Zamir and I published a paper describing the results of an experiment we conducted on the ultimatum game in changing environments.[4] In a stable and homogeneous society, norms of fairness will also be stable and unchanging. But in dynamic societies in which immigrants and people of differing cultural backgrounds mix, standards of fairness are created in a process of learning and constant adaptation; in such situations norms can change much more rapidly than we tend to imagine. To understand these dynamics, we brought together many players in our laboratory. Each player played the ultimatum game repeatedly, each time facing off against a different player. After playing about ten games against human players, some of the players in this experiment were paired off against virtual opponents—computer programs that we had created.

There were two types of virtual players. Virtual player of type A was programmed to make particularly low offers, between 13 percent to 16 percent when playing as the proposer, and to accept any offer above 16 percent when playing as the responder. Virtual player of type B was programmed to make generous offers between 45 percent to 50 percent as the proposer and to accept only offers that were above 45 percent when playing as the responder.

One group of human players in this experiment was paired off against virtual players of type A after playing ten games against a human opponent, while a second group of players was similarly paired off against virtual players of type B. The human players were not informed that at a certain point they would be playing against a computer program.

The experiment was conducted in Israel. In the first part of the experiment, in which human players faced human players in a series of ten games, the players played consistently with the norm of fairness typical of Israelis—the most common offer was slightly below 40 percent. But after a further ten to fifteen games against virtual players, the two groups adopted different norms of fairness. Human players playing against type A virtual players made offers that ranged between 20 percent to 40 percent while the players facing off against type B virtual players shifted to making offers that never fell below 50 percent.

These new norms were rapidly adopted under the pressures of two different forces. The human players who were paired off in the role of the responder against a proposing type A virtual player had to contend with very low offers that they initially rejected. With time, however, they were forced to accept those offers because continued rejection would have meant that they would have walked away from the experiment with very little in their pockets. Human players playing the role of the proposer who attempted to make outrageously low offers, down to 17 percent, when playing opposite a virtual responder of type A were surprised to discover that those offers were consistently accepted. That encouraged them to continue to experiment with low offers. Eventually most of their offers dropped down to very low levels. Similar dynamics, but in reverse, were observed among the human players playing against virtual players of type B. Offers that were even slightly less generous than an even division of the money were rejected and players were "educated" to offer only equal divisions.

We concluded from this experiment that norms of fairness can be quite fragile. A principled stand to reject any offer I regard as insultingly low can easily disappear if I see that almost all the offers I get are insultingly low. In fact, such offers will then cease to be insulting almost by definition.

The behavior of the proposers in the ultimatum game is consistent with assumptions of selfishness and rationality. The behavior of the responders, however, remains a bit of a mystery. Why would anyone in the role of accepting or rejecting an offer leave money on the table just to punish the other player for making an insultingly low offer, when the game is played only once and the two players will never see each other again? Robert Aumann suggested an interesting answer in a distinction he makes between "act rationality" and "rule rationality." According to this theory, limits to the cognitive resources available to us cause us to adopt simple behavioral rules that work well in most of the social interactions we encounter, but not necessarily in all of them. In other words, rather than planning out every little detail of our interactions, we settle for a pretty good plan and stick to it.

The rule of thumb that the responders to offers in the ultimatum game are using can be summarized as "never look like a sucker." Since most of the important social interactions we have in our lives are repeated interactions, sticking to this rule is efficient. In repeated interactions an expression of willingness to accept low offers will likely lead others to try to exploit us the next time we interact with them again. Rule rationality is often driven by emotions, especially what we have called rational emotions. The desire for revenge and punishment, a sense of insult versus a sense of honor, are all elementary mechanisms for creating optimal rules to be used in daily interactions that are similar to the ultimatum game.

This theory has recently been supported by the results of an important neuroeconomic study. Neuroeconomics is a new field of inquiry in economics that studies the brain activity that takes place in people when they make economic decisions.[5] Researchers in economics and psychology have increasingly been using magnetic imaging of the brain in recent years in order to map brain activity while decisions are being made. The specific regions of the brain that are being used at any given moment are identified using measurements of oxygen consumption.

In one study, fMRI devices measured the relative activity in different parts of the brains of subjects while they were playing the role of responders in the ultimatum game. Researchers discovered that extremely low offers triggered activity in parts of the brain associated with disgust and the vomit reflex. The sense of disgust that accompanies our reactions to insulting offers is possibly part of a mechanism that evolved to protect us from being exploited in repeated interactions.

In short, it seems, people are literally disgusted by unfair behavior. Do we really want to use reason to talk ourselves into accepting it?

PART II
On Trust and Generosity

7

ON STIGMAS AND GAMES OF TRUST

Why Did the Bees Commit Suicide?

TWO RESEARCHERS BASED IN WASHINGTON, D.C., PHILIP KEEFER AND
Stephen Knack, set out to ascertain the extent to which people trust
strangers in a research study that they published in a leading economics journal in 1997.[1] Thousands of individuals in dozens of
countries were asked to rate their trust in people they did not know
well, ranging from their car mechanic and primary care physician to
government officials responsible for public services. One of the more
interesting findings in that study was a strong correlation between
the trust people are willing to give to strangers and the GDP of the
country in which they live.[2] Countries with high levels of trust in
strangers have correspondingly higher GDPs. The study did not reveal a direct causal link between trust and economic development,
but subsequent research studies, some of them using laboratory experiments, have been able to uncover convincing underlying reasons
for the correlation.

Trust is an engine of cooperation between individuals. Cooperation, in turn, is an engine of economic growth and social welfare.

61

Trust cannot be sustained in a society without credibility, the behavioral trait that fosters trust. On the other hand, just as trust cannot survive for long without credibility, credibility is eventually destroyed without trust. If trust is virtually nonexistent in a social setting, then there is no point in trying to develop or sustain credibility; in that situation you are better off adopting selfish and unreliable behavior. Societies and nations can be in one of two equilibria: a "good" equilibrium in which individuals trust each other and behave in a reliable and cooperative manner toward others (justifying the trust), or a "bad" equilibrium in which individuals do not trust each other, with that lack of trust becoming self-justifying as people act without any sense of a need to be trustworthy or reliable. It is easy to guess, even without empirical data, which of these equilibria leads to greater economic growth.

Economists are divided on the question of whether these equilibria emerge from random processes or are dependent on initial conditions. If the first opinion is true, then the difference between contemporary Angola and Switzerland is due to long-ago random events that caused Angola to be trapped in a bad equilibrium, while Switzerland found itself in a good equilibrium. According to this view, there was once an equal chance that an alternative history could have developed leading Angolan society to look like Switzerland, while the Swiss would today be living like Angolans. Those holding the opposing opinion claim that there were initial conditions (such as a prevalence of natural resources, or a certain mix of cultures, etc.) that determined which country would be lucky enough to end up with a good equilibrium and which would find itself in a bad equilibrium. Those initial conditions could involve climate, geography, or cultural elements.

Of course, none of this matters if it is possible for a society to shift from one equilibrium to another (hopefully, from bad to good). Economic researchers are divided even more harshly on this matter, which has been labeled "convergence." Supporters of the convergence theory, who are apparently optimists by nature, claim that it is only a matter of time until Angola moves to a good equilibrium that will grant its citizens a living standard equal to that enjoyed by the Swiss. Their opponents claim that such equilibria are "ergodic" or "absorbing," meaning that it is difficult to move from one equilibrium to another (because the bad equilibrium "absorbs" the change rather than being overturned by it). The scenarios for going from a good equilibrium to a bad equilibrium are somewhat easier to imagine: food or water shortages, disease outbreaks, collapse of the government—any of these might precipitate a breakdown of a country's social order. But it seems to be especially difficult to go from a bad equilibrium to a good equilibrium. Imagine, for instance, that you are asked to move a large trunk from one room at your friend's apartment to another with the help of three other people. The trunk is so heavy that it can be lifted only with the full effort of the four of you. Following several failed attempts to lift the object, it would be pretty hard to get it moved. Each of you is likely to be suspicious about how much effort the others are exerting and whether they themselves believe that the job can be done. It would take quite a bit of discussion to get the job done, as by the time mistrust builds up it would require changing the behavior of the four of you to move to a better equilibrium. If at some stage you managed to pool your efforts together and get the trunk lifted, you'd be moving to the good equilibrium, but this new equilibrium is pretty fragile. It would take only one of you to change behavior (shirk a bit) for the object to fall and for the trust to collapse.

Although both camps in the dispute over convergence theory make use of extremely complex mathematical models, no resolution of the matter has yet been attained.

Research in economics differs in some ways from research in the natural sciences. Much contemporary economics research is theoretical, making use of mathematical models. This theoretical work is not dissimilar to theoretical work in physics, which also uses mathematical models. But unlike physics, where the ultimate test for determining whether a theoretical hypothesis is true is in the supportive empirical data, there are many economic theories that are widely accepted even without being tested empirically. In many cases there is simply no way to create the empirical results that could support or refute a given theory. How in the world could we use empirical tools to settle on the truth or falsehood of a theory that claims that within the next 1,000 years the living standards of Angola will converge to those in Switzerland?

This type of theoretical research is nevertheless very important in economics; human behavior is much too complex to describe precisely using mathematical models. Instead, the role of such models is more often the clarification of a claim or insight that could also be described without a model. Models in physics are the essence of the science; models in economics are tools. There are some fancy models in economics that show why a monopoly makes a higher profit compared with a firm that operates in a competitive market. These models provide many important insights, including ones that are relevant for policy making, but these models are far from describing the entire picture, and they are rather useless for purposes of making predictions about the economy.

In the 1990s, three American economics researchers suggested using a simple game called the trust game, suitable for laboratory experiments, for studying the extent of the trust and credibility people are willing to extend to others.[3] There are two players in the trust game: the first player (the proposer) is given a sum of money, say $100. This player can keep this money for himself or alternatively propose giving some of it to the second player (the receiver). For every dollar that the proposer gives the receiver, the experimenter adds two more dollars to the amount given to the receiver. For example, if the proposer grants $20 (out of the original $100) to the receiver, then the receiver will end up holding $60 (three times as much). At that point the receiver has the option of giving some of the money she is holding back to the proposer, as generously (or stingily) as she wishes.

Try to put yourself in the shoes of the players of this game and imagine what you would do. Your behavior as the proposer clearly depends on the amount of trust you are willing to put in the receiver.

If you choose to keep the entire original amount for yourself, you will go home with $100 and the receiver will go home empty handed. On the other hand, if you give her some of that money, which will then be multiplied by three, and she in turn gives you half of that tripled amount of money, then both of you will end up better off. If you are really daring and you give her the entire $100 sum, then she will hold $300 in her hands; if she then gives you half of that back, both of you will end up with $150, a tidy profit all around.

But the receiver has no incentive to share the money she gets, other than good will, a sense of generosity, or a sense of shame for acting in an ungrateful manner. You, as the proposer, find yourself faced with a dilemma. If we assume selfishness and rationality on the part of both players, then game theory would predict that the proposer will never offer a penny of what she initially gets to the receiver because she can be certain that the receiver will end up giving her nothing.

Like the ultimatum game, the trust game rapidly became one of the most prominently discussed games among behavioral economists. Unsurprisingly, from the start laboratory experiments on behavior in the trust game showed that proposers were typically willing to give a significant amount (usually about one third) of the money they had to receivers. Receivers, in turn, usually rewarded this generosity by returning to the proposers the original amount given to them by the proposers in addition to a small bonus.

The significance of the trust game, however, is not that it shows that people are willing to trust others to some extent but is its ability to measure and compare the extent of trust across different cultures. Several interesting experiments have been conducted relating to this.

To take one example, Uri Gneezy and Chaim Fershtman, two Israeli researchers, set out to study the effects that ethnic origins have on people.[4] They included in their experiment students at Tel Aviv University and Haifa University whose ethnic background—either European or Middle Eastern—could easily be identified by their family names. The participants in the experiment played the trust game via computer terminals, with the proposers located in Tel Aviv and the receivers in Haifa, about sixty miles away.

Each player was informed of the name of the player against whom he was playing. Players were paired in all the possible combinations: proposers of European background with receivers of Middle Eastern origin, proposers of Middle Eastern background with receivers of European origin, two European players, and two Middle Eastern players. The surprising and socially disappointing conclusion was that receivers with Middle Eastern family names were given significantly less than receivers with European names when proposers were called on to decide how much they were willing to give receivers. This discrimination against players of Middle Eastern background was mainly due to the behavior of European players, but Middle Eastern players also exhibited a degree of discrimination against players who shared their ethnic origin. Men tended to discriminate on this basis more than women. In other words, men systematically trusted players who had a European name more than they trusted players with a Middle Eastern name.

Discrimination, it turns out, is alive and kicking, as revealed by a simple experiment. We no longer see blatant discrimination around us because society strongly frowns on openly discriminatory behavior. But when we are far from the social spotlight, latent discrimination can rear its ugly head. In this experiment, the source of discriminatory behavior was an intuitive (perhaps even subconscious) feeling that many proposers had, leading them to believe that European receivers would be more likely to reward them for generosity than Middle Eastern receivers would be. Even Middle Eastern proposers apparently felt the same way about people in their own ethnic group, given the discrimination that they exhibited.

One might ask at this point if this intuitive feeling—that Middle Eastern players would be stingy in rewarding generosity—was justified by what the experiment revealed about the behavior of the receivers in the trust game. The answer is not at all. Regardless of their ethnic background, all the receivers tended to reward generous proposers to the same extent. In fact, Middle Eastern receivers were slightly more generous than European receivers!

How did the stigma against people of Middle Eastern background originate? In the previous chapter we mentioned the distinction that Robert Aumann proposed between rule rationality and act rationality. A rule rational action, as its name implies, is an action that is based on an instinctive rule that is favorable for us on average when we are involved in many different interactions over a lifetime, while an act rational action requires much more cognitive attention and is appropriate for a specific interaction.

Trust and mistrust are governed primarily by emotional rules. But while rules are effective in allowing us to reach quick decisions, they also have a major downside in their reliance on overgeneralization. The discrimination displayed in the experiment described above is an example of this type of misleading generalization. It stems from the perception that we should not trust people who are different or less well-off than us. While this might be a reasonable way of behaving under some circumstances, it could be detrimental to our interests in others. Such rules often form after just a very few instances in which we trusted the wrong people, and they are hard to modify. Indeed, they tend to survive for a long time even when they are proven to be harmful and wrong.

In this respect humans are not much different from bees, which also rely heavily on rules they find very hard to uproot. An interesting experiment on this was carried out several years ago in Germany with the aid of "artificial flowers." An artificial flower is a colored round box containing nectar that is very attractive to bees. The experimenters created a field of artificial flowers painted in different colors, yellow or blue. Nectar was placed in the yellow flowers, but the blue flowers were left empty.

A swarm of young bees was released over the artificial field. The bees immediately began to flit between the flowers. A bee landing on a yellow flower was able to fill up on nectar while the bees who found themselves on blue flowers were quickly disappointed and moved on to another flower. Over time the number of bees landing on blue flowers gradually decreased until all the bees learned to avoid the blue flowers and flew directly to the yellow flowers every time researchers released them over the artificial field.

At that point the experimenters switched the rules on the bees: they placed nectar in the blue flowers and left the yellow flowers empty. The expectation was that the bees would gradually learn that they should switch to the blue flowers and would desert the yellow ones. But no, that did not turn out to be the case. The bees stubbornly continued to visit only the yellow flowers, maintaining their previous behavioral pattern. Locked in a false stigma, they avoided the blue flowers despite repeated frustration in every visit to an empty yellow flower. This persisted even as the bees continually lost strength from lack of nourishment. Eventually the entire bee population died. In a sense, the bees committed suicide on the altar of the "stigma" they had applied against blue flowers.

The bee experiment teaches us about the perils of our unconscious biases but also suggests a way in which humans are able to resist them. As the trust game illustrates, our willingness to place ourselves in the hands of others can be changed by social conditions. As these experiments show, that kind of environment is only possible where emotions supersede pure, logical self-interest.

8

SELF-FULFILLING MISTRUST

I N 2001 I WAS APPOINTED PROFESSOR AT THE E UROPEAN U NIVERSITY Institute (EUI) in Florence, and I took up residence in that beautiful city. The EUI was founded by the European community for training the best minds on the continent in doctoral studies and academic research in the social sciences. Each member of the European Union is allotted a certain number of students that it can enroll at the institute, and as a result the institute is comprised of students of many different nationalities in equal proportion—a wonderful thing for the university, and also, coincidentally, a fantastic group to use for research. Most of the students speak at least three European languages and have lived in more than one country in the European Union. The EUI openly expresses its intention of being the ideological center of the European Union.

In late 2001 the German foreign minister, Joschka Fischer, convened the entire social sciences faculty of the EUI, about thirty professors, and charged us with the strange task of drafting a constitution for "the United States of Europe." When I was subsequently awarded a large research grant from the European Union, I chose, along with several colleagues, to use part of the money to study trust and trustworthiness in the European context.[1] We decided to conduct an experiment using a game based on a "market of favors."

Students from different parts of Europe who had just arrived at the EUI (and therefore had not had any time to become acquainted with each other) were the subjects of the experiment. They were divided into groups, each with five students. The members of each group did not see each other directly; all their interactions took place through computer monitors.

At the start of the experiment a brief description of each participant was circulated among the members of the group. The most important detail for us (the experimenters) was the participant's country of origin, but we also included age, academic interests, and other minor details in the circulated descriptions. Each participant was then given fifty euros and instructed that he could give any amount of money from that sum to any other member of the group. As in the trust game, any money that was thus transferred from one participant to another was tripled. The recipients of this initial generosity were given an opportunity to repay the proposers who had given them the money, in any amount that they saw fit, again just as in the trust game.

This was repeated in each group for six rounds. The resulting effect was the creation of a dynamic market of favors in which individuals chose others as recipients of their largesse, with the expectation that they would be rewarded for this generosity, either in the same round or in a later round.

Our aim was to compare the extent to which people are willing to trust individuals from northern Europe compared to those from southern Europe. For the purpose of this study we defined northern Europe to include Denmark, Sweden, Finland, Great Britain, Germany, the Netherlands, and Belgium. We considered Italy, Spain, Greece, Portugal, and France (most of whose citizens live in the southern part of that country) to be southern European countries. Not coincidentally, the geographic line dividing northern Europe from southern Europe is also the cultural dividing line between Latin culture and Anglo-Germanic culture.

Given the backgrounds of the subjects in this experiment, who were all young intellectuals with resumes that included a significant amount of international and multicultural interactions, one might expect that country of origin would have no effect on the trust levels exhibited in the game. But that turned out to be a wrong assumption. Southern Europeans were significantly discriminated against relative to northern Europeans. Northern Europeans exhibited distrust of southern Europeans. The discrimination was evident both in the identities of the partners chosen to be recipients of money and in the amounts of money given to them by the other players. Southern Europeans were chosen relatively infrequently, and when they were chosen, they received less money compared to northern Europeans.

The dynamic aspect of the game gave us an opportunity to follow how discrimination emerged in round after round of the game. We had expected expressions of discrimination to diminish as the game progressed, but to our surprise the opposite happened. A careful analysis of the data revealed the secret of growing discrimination: in the first round there was some expression of mistrust against southern Europeans, but it was small and marginal. The objects of this mistrust then responded with their own measure of distrust in the second round; mistrust naturally fosters a mistrustful reaction. This was interpreted as justifying discrimination against the southerners, leading to even greater discrimination against them in the next round, further entrenching mistrust, and so on in a growing spiral of discrimination and mistrust. A small and marginal initial grain of unjustified discrimination ballooned out of proportion before our eyes.

The small mistrust exhibited at the start of the game was a self-fulfilling prophecy that by the end of the game became full-blown discrimination. We concluded that if young, sophisticated intellectuals enrolled in the elite EUI could act this way, the phenomenon should be prevalent throughout Europe.

Publishing the paper proved difficult because some reviewers, unfairly in my view, regarded it as being provocatively accusatory. To his credit, the editor of the most important journal in behavioral economics, a Spaniard from a German family, recognized the importance of the paper and agreed to publish it.

Both southern Europeans and northern Europeans bore some measure of responsibility for the growing mistrust that emerged in our experiment. Many interpersonal communication failures apparently stem from such self-fulfilling mistrust. An employer expressing lack of confidence in the abilities of an employee limits that employee's chances of successes. If that employee then fails as a result, the employer will feel that his initial expectations were confirmed. On the other hand, an employee who from the start expects any job success to be summarily dismissed by his or her employers is inviting the very lack of credit and respect that he or she expected. Fear of being hurt or disappointed in a romantic relationship can itself doom the relationship.

9

CULTURAL DIFFERENCES, PALESTINIAN GENEROSITY, AND RUTH'S MYSTERIOUS DISAPPEARANCE

IN 2008, REINHARD SELTEN AND I RECEIVED A RESEARCH GRANT FROM the German Science Foundation to conduct a laboratory study of ethnocentrism, the judging of people in other societies solely based on one's own cultural norms. Along with Palestinian colleagues from Bethlehem University and Al-Quds University, we conducted two experiments involving Germans, Israelis, and Palestinians playing the trust game. Recall that the trust game involves two players, a sender and a receiver. In the first stage, the sender is endowed with an amount of money by the experimenter, from which he can transfer any part to the receiver. For any dollar transferred to him from the first player, the receiver gets two additional dollars from the experimenter. In the second stage the receiver can transfer back any part of his income to the sender.

In the first research experiment we had players of each nationality play the trust game while facing off solely against players from the same nationality: at Bonn University, German players played against German players; at Hebrew University in Jerusalem, Israelis played Israelis; and at Al-Quds University in Jerusalem, Palestinians played Palestinians.

Having the players initially play only others of the same nationality enabled us to establish a baseline of trust within each group separately. There were significant differences between the groups. Palestinians exhibited the greatest amount of trust, offering, on average, 66 percent of the money they had to other players. In contrast, Israelis were the least trusting, making, on average, offers of only 36 percent. The German group was in the middle of this ranking, with an average offer of 50 percent.

The players in this experiment played the standard trust game, but we also asked players in the role of receivers to predict beforehand the offer they expected to get from the proposers. On average, these guesses turned out to be surprisingly accurate, within each group. The behavior of the proposers was highly correlated with the expectations of the receivers. Palestinian receivers at Al-Quds were not in the least surprised at the generous offers they got from their fellow students, while the Israeli students at Hebrew University were not surprised at the low offers that their Israeli counterparts tended to give. Apparently each culture has its own internal norms of trust, which are well known to the individuals living within those cultures.

But how exactly did the participants in this experiment learn the norms prevalent in their own cultures so well that they could guess the behavior of the proposers almost exactly? Each of our participants was playing the trust game for the first time ever and had no prior experience in the particulars of that game. We all, however, throughout our lives and on a daily basis participate in interactive situations that may not be precisely identical to the trust game but resemble it in many ways. Our experiences in such situations occur so frequently and are so much more significant than any single play of the trust game that the cultural norms with respect to trust and generosity prevalent in our environments are etched within our intuitions. Having such intuitions is critical for social success. In fact, it may be more important than the ability to analyze situations in which we find ourselves.

In the second experiment we had players from each culture playing against players from different cultures. All the possible cultural pairings were effected: we had Israelis paired with Palestinians, Israelis paired with Germans, Palestinians paired with Germans, Germans paired with Germans, Israelis paired with Israelis, and Palestinians paired with Palestinians. The experiment was conducted simultaneously in Bonn, Jerusalem, and the West Bank via electronic communication. Each player was informed of the nationality of the player with whom he was paired.

Ethnocentrism emerged in full force in this experiment. Players in the role of receivers made the same initial predictions regarding proposer behavior as they had in the first experiment when they were paired with proposers from their own cultures. For example, Palestinians continued to guess that the average offer would be 66 percent even when they were paired with Israeli proposers. Israelis, who were used to receiving far less generosity from their co-nationals (37 percent offers), expected that same low level of trust when paired with Palestinian or German proposers.

What about the proposers? They also conformed to the actions of their own cultures, making more or less the same offers in both the first and second experiments, whether they were paired with players from their own culture or from a different culture. Because the proposers' offers were the same regardless of the identities of the receivers against whom they were paired, we can conclude that there was no significant discrimination based on nationality.

The absence of overt discrimination here may seem encouraging, but peering a bit further under the surface reveals a less cheerful picture. The innocent behavior of the players in this experiment involved a level of ethnocentrism that carries with it potentially dramatic and even tragic effects. This became clear when Israeli proposers were matched with Palestinian receivers. Israeli norms, as expressed in the first experiment, were for proposers to offer very little (about 36 percent on average). Israeli proposers consistently made such low offers whether the receivers were Israelis or Palestinians.

Palestinian norms, however, involved much higher standards of proposer offers (about 66 percent on average). Ethnocentrism led Palestinian receivers to expect Israeli proposers to make offers in line with that Palestinian norm. They were therefore inevitably disappointed when they saw the offers the Israeli proposers were putting forward. In questionnaires distributed to the participants at the end of the experiment, the Palestinian players explained the gap between what they expected from the Israeli players and what they actually received as an expression of Israeli discrimination against Palestinians. They did not even consider the possibility that the gap was actually caused by different behavioral norms and that the Israelis made the same low offers to their fellow Israeli players. Many of the most dangerous elements of ethnocentrism are caused by our failure to recognize that different cultural norms can exist at all.

An opposite, positive effect was registered when Palestinians were in the role of proposers and Israelis were receivers. In that case, ethnocentrism caused Israeli receivers to expect low offers from the Palestinian proposers, on the order of 36 percent, just as they received from Israeli proposers. They were pleasantly surprised when the average offer they received, 66 percent, was nearly double their expectations. As shown by their questionnaire answers, the Israeli players did not consider the possibility that what they had witnessed was the Palestinians simply behaving according to the norms of their culture. Many of the Israelis considered the offers they received to be an inexplicable positive gesture from the Palestinian players to the Israeli players.

Ethnocentrism exists wherever differences in cultural behavior are encountered. What we haven't yet considered here is why the Palestinian norm is to offer so much more than Israelis and Germans in the trust game. Why should they have so much trust in others in this game? The high offers the Palestinians persisted in giving Israelis and Germans shows that special feelings of identification with the other students in their universities or Palestinian solidarity cannot be explanations for the phenomena.

I cannot claim to have a convincing explanation. All I can give are speculations gleaned from long conversations with the Palestinian colleagues who conducted the experiment with me, and chiefly Mohammed Djani of Al-Quds University. My colleagues ascribe the generous offers made by Palestinians in the trust game to the differences in the relative importance given to collectivism versus individualism in Palestinian culture. Individualism is still considered to be disgraceful in Palestinian society because individualism conflicts with traditional and religious values. The Israeli-Palestinian conflict may also make Palestinians wary of over-individualism.

Failure to reciprocate an act of generosity is considered much more contemptible in Palestinian society than in Western cultures. This leads Palestinian proposers in the trust game to be more generous toward receivers. Surprisingly, ethnocentrism causes Palestinian proposers to expect receivers from other cultures to respond in the same way as Palestinian receivers would, when in fact egoistic behavior is much more legitimate and prevalent in Western cultures.

Spreading Internet usage and economic globalization are accelerating the pace of intercultural interactions. Ethnocentrism might disappear within less than a century, not because we will learn to appreciate the behaviors of people in other cultures but because differences between cultures will nearly disappear. A single canonical model of behavior will be shared by most of humanity. Whoever fails to act in accordance with that canonical model will simply fail to survive, economically and socially. Until that process is completed, however, success will come more easily to those who are aware of their own ethnocentrism and manage to adapt their behavior to the social environment in which they find themselves, even if that means changing what they had been used to previously.

This is especially important when interactions involve negotiations, both commercial and political. As Raymond Cohen argues in his book, *Culture and Conflict in Egyptian-Israeli Relations*, negotiations break down often because of ethnocentrism rather than substantive differences between the parties. That is what has doomed many of the attempts to bring about a peace agreement between Israelis and Palestinians. Getting to an agreement will require more than just getting negotiators to overcome their ethnocentrism and imagine themselves in the place of their interlocutors. Most of the populations of both nations will need to overcome their ethnocentrism as well. Without broad popular support among Israelis and Palestinians, no agreement will ever be implementable.

My colleagues in the German-Israeli-Palestinian experiment and I conducted another experiment, with the goal of improving our understanding of cultural differences in economic interactions. This experiment involved a new two-way trust game, with two variations.

We called the first version of the new game we developed the "giving" version. The game is played as follows: each of the two players in the game is initially given an equal amount of money (such as $100). Each player then decides how much of this sum she is willing to give to the other player. As in the standard trust game, the experimenter gives a receiving player two dollars for each dollar that she got from the other player.

The two players in this game make their decisions on how much to offer the other player simultaneously, without knowing how much the other player is offering. The total amount of money that a player has at the end of the play of this game is therefore the amount that she kept without offering the other player, plus three times the amount received from the other player. For example, if the first player gave the second player $30 while receiving $20 from the second player then his sum total at the end of the game is $70 + $60 = $130.

The second version of this game is the "taking" version. The game is played as follows: as before, each player is initially given $100. Each player then declares how much money she will take as a "cut" from the original amount of the other player. The experimenter gives each player two dollars for every dollar that remains in her possession after the cut to the other player has been transferred. The total amount that a player has at the end of the play of the game is the sum of what he or she took from the other player plus three times what remained in her possession after the other player took a cut of the original amount. For example, if the first player took a cut of $70 from the second player, while the second player took a cut of $80 from the first player (leaving him with $20), then the first player ends this play of the taking game with a sum of $70 + $60 = $130.

From a strategic perspective, both of these games are identical: the taking game can be reformulated as a two-stage game as follows: in the first stage each of the players takes the entire amount of $100 from the other player and then in the second stage the two of them proceed to play the giving game. But that transfer of $100 in the first stage is meaningless because each player is left with $100 at the end of that stage.

Note that if both players are selfishly rational, thinking only of their own private gain, then neither player will offer the other player even a penny in the giving game, leaving both players at the end with the $100 they started with. In the taking game, if both players are selfishly rational, then they will each take $100 from the other player, leaving both of them at the end with the $100 they started with. These behaviors are exactly the Nash equilibria of the respective games. Given the strategic equivalence between the two games, self-interested rationality dictates that we should expect similar behavior in both games, with nearly identical end results; the games differ from each other only in how they are described. But the results of the experiment were very different. Players behaved significantly differently when playing the giving game compared to their behavior in the taking game. Importantly, these behavioral differences were culturally specific.

My colleagues compared the behaviors of Israeli, Palestinian, and Chinese players in the experiment using the giving and taking games. Each pair of players always involved players from the same culture. Israelis gave relatively small amounts in the giving game, but took large amounts for themselves in the taking game. Palestinians gave relatively large amounts in the giving game but also took large amounts for themselves in the taking game. The Chinese players gave relatively small amounts in the giving game and also took small amounts for themselves in the taking game.

The players in the different cultures had different behavioral characteristics. Israelis placed a great deal of importance on their personal gains. Their behavior was the closest to the behavior predicted under Nash equilibrium. Do not, however, conclude from this that Israelis are generally more selfish than individuals from the other two cultures. We will return to this important point later.

Palestinians, in contrast, were very generous in the giving game while being selfish in the taking game. This hints that Palestinians emphasize nonmonetary considerations, such as expectations of reciprocity, in their decision making. Their behavior was influenced by what they expected others to do, which in turn depended on the detailed way each of the two variations of the game is described. Describing the game in terms that emphasize giving leads to expectations that others will be generous, encouraging everyone to be generous in accordance with the general norm. On the other hand, describing the game as one in which players decide how much to take gives rise to egotistical expectations, which then turn out to be self-fulfilling.

The Chinese players showed a respect for property and a striving to avoid overly generous actions on the one hand and causing harm to others on the other hand. They gave in moderation and in turn took in moderation, preferring as much as possible to end the game with the same amount of money as they received at the start of the game. The behavior of the Chinese players reminds me of stories my father-in-law told about his experiences as a soldier in the Soviet Red Army during the Second World War. He would always end each story by repeating the same sentence, which summarized his main insight from the war and possibly the secret of his survival: "Never volunteer and never refuse orders."

I was not surprised by the behavior of the Israeli players in the taking game. I have had many discussions with colleagues about the consistent phenomenon of Israelis exhibiting much more competitiveness and utilitarian self-interest than individuals from other countries in experimental games. The phenomenon has been noted in a broad range of such games, including the ultimatum game, the Prisoner's Dilemma, the dictator game, and the trust game. This is particularly disturbing to Israeli researchers, who often express discomfort and embarrassment at international conferences when they present the results of experiments conducted in Israel, along with concern that highlighting this phenomenon could serve the interests of peddlers of base anti-Semitic slurs. But do the experimental results really point to a deep-seated Israeli tendency toward competitiveness, excessive greediness, and callous lack of concern for others? I don't think so.

I believe that the root of this phenomenon is in the dissonance that exists between Israeli individualism on the one hand and the special place that giving and solidarity have in Israeli society on the other hand. In times of crisis, Israelis exhibit a ready willingness to cast aside personal interests and voluntarily join together for mutual assistance that is rare even by the most demanding international standards.

Israeli society could not have survived a bitter one-hundred-year-long conflict if it was composed entirely of greedy individuals, each looking out only for his or her immediate narrow interests. But that only sharpens the question: Why don't these examples of overwhelming solidarity and mutual concern find expression in laboratory experiments?

The answer, I believe, lies in the high value Israeli society simultaneously places both on solidarity in the face of crises and on individualism and success. It is this combination that is the secret behind the economic, scientific, and technological successes Israel has experienced. Solidarity and readiness to pitch in for the common good exact a price that Israelis as individuals are willing to pay when social or security crises erupt. But in calmer situations the average Israeli seeks to express other values, such as competitiveness and success, as a respite from the heavy burden of solidarity. To balance out that burden, he allows himself the natural right to behave in a more individualistically instrumental way than his European or American colleagues, who are less often called upon to rally to solidarity. The following story sheds some light on the interactions between solidarity and rational emotions in Israel.

The summer of 2006 was unusually hot in Israel. For several very long weeks, a bitter military conflict raged between Israeli forces and Hezbollah in Lebanon. Hezbollah managed to strike at Israeli civilian targets harder than they had been hit since 1948.

In the middle of that hard-fought war I received an e-mail from Doris, an acquaintance of mine, whose daughter Ruth had come for a summer visit to Israel just before the war began. I've known Doris Ericson and her husband Larry since 1990, when I was a postdoctoral fellow at the University of Pittsburgh. My wife and I often spent weekends with them and their daughter Ruth, who was ten years old at the time. Ruth was a true wonder child. She had mastered three languages and several musical instruments, and did not hesitate to correct any of us if we made the smallest mistake, whether that was misquoting the price of an item in the supermarket or misunderstanding the concept of a quantum bit of information in computer science. After we returned to Israel, we naturally saw the Ericsons less often. In 2006 we had not seen Ruth, who had by then turned twenty-seven, in nearly seventeen years.

On that unusual Friday, while a war was raging, Ruth was supposed to come to our place for dinner. At around 7 p.m. she called from the center of Jerusalem to ask me to explain to the driver of the taxi she was in how to get to our house, and to do so quickly because the battery on her mobile phone was nearly empty.

The trip from the center of Jerusalem to our house normally takes about twenty minutes. At 7:45, when Ruth still hadn't arrived, I rang her telephone number to ask if everything was all right. There was no reply. As my level of concern for Ruth began rising, my wife Atalia tried to soothe my worries. "Of course she isn't answering." said Atalia, "Don't forget that her phone battery is dead." But by 8:15, when Ruth still hadn't arrived, even Atalia could no longer hide her anxiety.

At 9 p.m. we couldn't stand anymore: we decided to call the police. Instead of being told by the policeman on the other end of the line that I was being a nuisance by contacting the police for no good reason in the middle of a war (which is actually what I had hoped I would be told), I was extremely alarmed by what he said. "What you are reporting, Mr. Winter," said the policeman, "sounds very serious. You should have called us earlier." Within less than half an hour a police patrol car was at our house, with the police force already engaged in an effort to identify the exact location from which Ruth had last made contact with us by phone, in the hope of tracing where she might be.

"Mr. Winter," asked a policeman, "what can you tell us about the taxi driver's reaction to the route you suggested to him?"

"He simply said something like 'No problem, I understand.'"

Many more questions followed: "Did he respond in the middle of your explanation or only when you had finished? Did he have any identifiable accent?"

When all attempts at locating Ruth using her mobile telephone came up empty, the police sent a patrol car to the area from which she had last contacted us. They asked us to supply a detailed description of Ruth's physical appearance. We told them that we were unable to do so.

"What do you mean?" asked one of the policemen angrily. "Is there someone named Ruth who was supposed to visit you this evening or not?"

When Atalia and I tried to reconstruct what Ruth looked like as a ten-year-old girl, the policemen lost their patience. "Isn't there anyone who knows Ruth well enough to give us a description of what she looks like today? What about her parents?"

As I mumbled that her parents certainly know what Ruth looks like, the sentence I was afraid I would hear followed: "Then get Ruth's mother on the phone."

The handful of minutes during which I spoke on the phone with Doris, who was at her home in Pittsburgh, were the most difficult I experienced that evening.

"Doris," I said, "please listen—Ruth hasn't shown up."

A heart-stopping silence fell over our conversation. I tried to calm both of us. "Doris, it's only been two or three hours since she was supposed to arrive. Maybe Ruth decided at the last moment that instead of a boring evening with friends of her parents she would rather take a taxi to a more interesting place. Maybe she is meeting a girlfriend or a boyfriend . . . "

"No, Eyal," replied Doris. "She doesn't know anyone there other than you and Atalia. I don't know what to do. Larry is at a conference in California, and I am alone at home. You must help me."

The conversation with Doris left me and Atalia feeling even more anxious than before. I decided to make one more attempt to contact Ruth's mobile telephone before turning again to the policemen in my living room. To my surprise, this time I heard Ruth's voice answering.

"Ruth, is it you? Where are you??"

"Eyal," replied Ruth, "I'm having dinner at your house."

"No, Ruth, where are you right now?"

"I'm at your house now," insisted Ruth.

At that point the police interrupted: "Tell her we are sending a patrol car straight away. Just get her to tell us where she is."

"I think she is confused," I stammered.

"Of course she is confused!" barked a policeman. "She's apparently been attacked. Try to get her to describe what she sees around her."

I handed the phone to Atalia, who I felt would be better at this task, as she is a trained psychologist.

"Hi Ruth, this is Atalia. Tell me exactly what you see around you."

"I see several people sitting around a long table. You're in the kitchen now, right?"

"Yes, of course," replied Atalia. "Could you please let me speak with one of the people at the table?"

"Sure, just a moment."

Within seconds a deep male voice could be heard speaking into the phone. It turned out to be a neighbor from only a few houses away. An irresponsible taxi driver had brought Ruth to the wrong address. That, along with some distracted thinking on Ruth's part and some over-the-top hospitality on the part of the neighbors (who had just moved into the neighborhood), had led to a comedy of errors that we did not in the least find amusing.

If my neighbors' behavior seems strange to you, keep in mind that this happened in the middle of a war. Atalia and I, like many others, had invited people from the north of the country, which was being attacked by rockets, to come stay in our town far from the range of the rockets. Our neighbors had innocently assumed that the young woman who showed up at their doorstep was yet another refugee from the north seeking temporary respite from the war. When they opened the door of their house in such a welcoming manner, Ruth was convinced that she had found the home of her parent's friends. She immediately slipped off her shoes, walked in with a broad smile and hugged every person standing there.

The others sitting at the table at my neighbor's house thought there was something a bit strange in the behavior of this uninvited guest but made an effort to avoid embarrassing her or making her feel unwanted.

By this point Atalia understood that Ruth had mistakenly found her way to a nearby house. She excitedly ran out to bring Ruth to our house, not even stopping first to put on her shoes. When my neighbors' eldest son saw Atalia running toward his house barefoot he shouted "Mom, there's another crazy lady coming here!"

For several hours, Ruth found herself welcomed as the guest of a family she did not know, with neither her nor her hosts considering the possibility that a mistake had been made, because they did not share the same cultural norms regarding the behavior expected of guests and hosts. Ruth did not imagine that she had arrived at the wrong house, because in the cultural norm with which she was familiar (that of the United States) it is inconceivable for a complete stranger walking into a random house to be as warmly welcomed as Ruth was. In judging the situation using her cultural norms, she fell into the trap of ethnocentrism.

For Ruth's hosts, a religious Israeli family, having a stranger as a guest for a Friday night meal is the most natural thing in the world, especially during a war in which many people had fled and were still fleeing from parts of the country threatened by rockets. They expected an individual under those circumstances to feel comfortable seeking a house and a meal. The warmth with which they greeted Ruth seemed self-evidently natural to them, and they strove to do everything to maintain her comfort level and avoid raising any subject that might cause her embarrassment. They too fell into a trap of ethnocentrism, albeit an ethnocentrism trap that was pleasant for those who fell into it (and rather unpleasant for some of us outside it).

In any event, after that evening I never again felt a need to apologize when Israelis appeared to be selfish and egoistical in laboratory experiments.

The story of Ruth's disappearance provides us with yet another important lesson on selfishness and generosity: Why did my neighbors allow a complete stranger to invade the intimacy of their weekend dinner? Would they have acted in a similar manner in time of peace? Clearly the generosity they displayed had a lot to do with the existing state of war in Israel at the time.

The tension between solidarity and individualism that is so prominent in the Israeli society exists, though in a more moderate form, in every society. At times of crisis such as wars or natural disasters, people will crave solidarity and will despise competitiveness and selfishness. But as soon as the threat is gone, these feelings will be replaced by an increased appetite for individualism and self-interest. Bourbon Street during Hurricane Katrina and Wall Street during a stock market boom are two different planets in terms of the way we think about individualism and solidarity.

People seem to be less selfish and more generous to fellow group members when the group is exposed to an outside threat. We term this behavior "solidarity," and it is crucial to the survival of societies.

Solidarity as a form of collective emotion is the topic of our next chapter.

10

COLLECTIVE EMOTIONS AND UNCLE WALTER'S TRAUMA

In the summer of 1933, only a few months after Adolf Hitler became dictator of Germany, Walter Lazar happened to walk straight into a large Nazi rally in the city center of Königsberg, in what was then Prussia, an eastern province of Germany. Walter, my grandmother Jenny's brother, was exactly the sort of liberal, cosmopolitan Jew that the Nazi regime regarded as the very embodiment of evil. Seeing the rally forming in front of his eyes, his first instinct was to flee the area as fast as humanly possible. But curiosity got the better of him. Instead of running away, he slowly worked his way into the heart of the rally. His Aryan appearance belied his Jewish ethnicity, and the people around him accepted his presence as natural.

Eventually, Hitler himself appeared on stage to deliver one of his characteristically fiery speeches, gesticulating wildly with his arms and screaming himself hoarse to whip the crowd into a frenzy. His sentences were frequently punctuated by chants of "Sieg Heil!" uttered by the thick throng, followed by total silence as the crowd eagerly awaited the Führer's next words. At first Walter took in what was happening around him with stunned disbelief. But then a strange feeling slowly took hold of him. When the rally sang the Nazi Party anthem, Walter joined in, mumbling the words to the song. Not long after that he suddenly noticed that he was actually getting swept up in the ecstatic crowd's powerful emotions. Along with everyone around him, he was also shouting "Sieg Heil" and applauding Hitler's every word!

Coming to his senses, he covered his face in shame and fled to his sister's home, not far away. My father, little Hans, Grandmother Jenny's son, was age twelve at the time. He recalled Uncle Walter's appearance that day for the rest of his life. Opening the front door, Hans noticed his uncle was as white as a sheet, drenched in sweat. Jenny was so alarmed that she took hold of the telephone to call a doctor, but Walter persuaded her that was not needed. He collapsed onto a sofa, crying bitterly. "It was like witchcraft. How could I have joined in singing the Nazi anthem and hailing Hitler?"

Walter's story, in fact, is not especially unusual. It is a historical fact that within several months of Hitler's takeover of Germany, many formerly committed members of the social democratic and communist political parties participated in massive Nazi Party rallies with great enthusiasm. An ecstatic group of people acting in unison like a single organism can have a powerful emotional influence on us, at an almost suggestive level. This stems largely from an ancient need we all have to belong to a group.

The evolutionary advantage that comes with belonging to a group is very clear. Being a member of a group gives an individual much greater security in the face of threats from dangers and enemies, along with better access to vital resources.

Several psychology experiments have shown that the human need to belong to a group is so significant that it exists even in abstract and context-free situations. Subjects who were divided into two color-coded groups (blues and greens) and played the trust game described in a previous chapter tended to be more generous to members of their own color group than with members of the "other" group, despite the fact that the color assignments had nothing to do with the game itself. The mechanism that creates and preserves group cohesion is at root an emotional mechanism eliciting collective emotions.

Scientific, technological, and artistic developments are primarily cognitive and emotional phenomena operating at the level of individuals. But the social history of humanity has been dictated mainly by collective emotions. Wars and treaties, along with great revolutions and sweeping political and economic changes, are largely driven by such emotions.

My late colleague and friend Gary Bornstein devoted many of his research efforts to studying the way that tensions between groups strengthen cooperation within groups. In two of those research studies in which I participated, we had subjects play slight variations of the Prisoner's Dilemma.[1,2] The game was played with two pairs of players, instead of a simple one-on-one match. Each pair of players separately played the one-on-one Prisoner's Dilemma and received the resulting payoff, but if one of the pairs ended up with a joint payoff greater than the joint payoff of the other pair, we gave each member of the "winning" pair a small extra bonus.

In contrast to individual emotions, which formed the subject of previous chapters, collective emotions enable some individuals to correlate their mental states. Correlated mental states can be expressed, for example, in a desire to out-compete rival groups. This explains why even in cases in which the winning payoff to each individual is small and may appear to be insufficient to justify cooperation, correlated mental states can motivate a great deal of group cooperation.

The extent of cooperation that was revealed in our color-coded group experiments was astounding, far more than the cooperation usually seen in straightforward one-on-one versions of the Prisoner's Dilemma (without groups). This cooperation came from a strong desire to out-compete the other group, even though the individual payoffs for cooperation were identical to those in the usual Prisoner's Dilemma. The cooperation that group identification prompted, however, enabled the subjects in these experiments to gain much higher payoffs than if they had played the usual Prisoner's Dilemma.

This simple simulation indicates how collective emotions enable individuals to improve their material conditions, bolstering their chances of survival. This clearly translates into an evolutionary advantage at the individual level. In the above example, correlated and committed collective emotions were elicited between only two cooperating partners in each color-coded group. But imagine a situation in which rival gangs are facing off against each other, ready for violent action. That is a classic example of collective emotions being generated. Each gang member is filled with empathy for his own gang mates, to the point of willingness to risk his life to save one of them or avenge the gang's honor. In parallel to all this, hatred for the members of the rival gang boils within him. These collective emotions have effects both in feelings gang members have toward each other and toward the members of the other gang. They generate internal commitments to fight for the group and to threaten the group's rivals. A group that can fire up these collective emotions among its members gains an advantage over other groups. This increases the group's chances of survival.

The human ability to coordinate emotions and turn them into a powerful force apparently has ancient evolutionary roots. Whoever failed to join in with the collective, either of his or her own choice or due to banishment, suffered significantly diminished chances of survival relative to those who were committed to the group. In fact, collective emotions can be identified in many mammals and birds; they are not limited to human beings alone.

The interested reader can watch an astounding video on YouTube titled "Battle at Kruger." The video was filmed by a group of tourists at Kruger National Park in South Africa. In its initial moments, a herd of buffalo is shown calmly meandering along a pastoral path on a river bank. Suddenly, a pride of lions seeking easy prey appear out of nowhere, focusing their attention on a young buffalo calf toddling between the legs of its mother.

After a brief but chilling chase, the lions manage to scatter the frightened herd. The poor calf, unable to keep up with the herd on its weak legs, is increasingly isolated, just as the lions intended. Alone in the open, the calf is readily trapped in the jaws of one of the lions, and is being dragged toward the river, seemingly doomed to be drowned and then eaten.

But even a baby buffalo is too big to be taken down easily to its death. The calf exhibits remarkable tenacity, fighting stubbornly for its life. As if this weren't enough, in the middle of this mighty struggle on the banks of the river a crocodile abruptly emerges from the waters, grabbing one of the calf's legs between its powerful teeth. The surprised lions do not give up, now attempting to pull the ailing calf away from the river while the crocodile just as doggedly pulls in the opposite direction, toward the waters.

The lions, who as a group are stronger than the crocodile, win the tug-of-war. Now the calf is gripped tightly in the jaws of three lions, its sad fate all but sealed. But then the most astonishing thing happens. The buffalo herd that had previously fled in the face of the lions comes storming back in a determined and angry march. Within seconds a large group of buffalo surrounds the lions clutching the calf, while another group of buffalo threateningly charges the rest of the lions to chase them away.

The ring around the three remaining lions closes in on them menacingly, until the frightened lions let go of their intended prey and flee for their lives. The buffalo calf, left injured on the ground by the lions, calmly rises to its feet as if nothing had happened and rejoins the herd.

This incredible Battle at Kruger shows how herbivorous buffalo can defeat a pride of lions, the most fearsome carnivores in the safari reserve, by harnessing the numerical power and cooperation inherent in a coordinated herd. The tourists who filmed the battle could not contain themselves. Their emotions overcame them, and they can be heard encouraging and cheering on the thundering buffalo herd in the video. The situation looked like such a human drama that the people seeing it unfold before their eyes could not help but be swept away themselves in a powerful collective emotion of identification. I believe that every person watching the video feels that same surge of collective emotion. It can be seen at:

I previously noted that collective emotions can sometimes be stronger than individual emotions. One reason is that in many social situations these two types of emotions become involved in a feedback mechanism with each other. In many religions, the devout congregate together for prayers in churches, mosques, synagogues, and so forth, not for the sake of the gathering itself but to create an environment in which the emotional power of prayer is magnified. Fans at football games become energized again and again by the other fans around them, with their energy then further feeding into the excitement of those around them, in a feedback loop. Teenage girls who physically swoon in the presence of a musical idol such as Justin Bieber almost always do so only when they are together in a group. In a one-on-one meeting with the same idol, they are more likely to react in a more composed manner.

There have been times in many societies in which political and ideological passions have been known to rend apart families, with spouses or children and parents refusing to speak to each other because of the contrasting opinions they hold with regard to divisive issues. When the passions cool after some time has passed, those involved often say they don't understand what came over them. How could they have reacted so extremely to a question that in retrospect seems minor? Those reactions, however, did not arise due to differing intellectual analyses of political questions alone. They were accompanied by collective emotions that included group identification, in this case identification with one ideological group in contrast to another.

Collective emotions often require the existence of an opposing group playing the role of a competitor or a source of threat. Preserving our collective "we" requires a collective "they." The greater the conflict between "us" and "them," the greater our collective identification with each other, making it easier for us to act as a group.

This is a familiar pattern of behavior for many of us. Under emergency conditions, such as when a hurricane is approaching, you will see people pitching in together for the protection of the collective. They will often be very generous toward each other. If the storm blows back to sea and the threat evaporates, people go back to concentrating mainly on their own personal lives. Patriotism and flag-waving are most prominent when there is a perceived enemy threatening the country. We may be very critical of our government, but then in the presence of a foreigner suddenly find ourselves passionately defending it and what it represents.

This is not restricted to Western nations and cultures alone. My friend Yoshi Seijo Matsuoka is the scion of a famous Japanese samurai family. One of his ancestors was the samurai warlord of the city of Osaka in its war in the seventeenth century against Edo (as Tokyo was then called). Tokyo is today the capital of Japan largely due to the outcome of that war.

After a devastating earthquake and subsequent tsunami struck Japan in 2011, I called Yoshi to check that he and his family were safe. We talked at length, not having seen each other for a long time. At a certain point I asked Yoshi about the criticism that many Japanese citizens were expressing against their government, accusing it of not organizing sufficiently quickly to care for the survivors of the catastrophe. Yoshi briefly replied to my question, but then almost angrily accused the Israeli government of acting in a "cowardly and ungrateful" manner when it hastily evacuated embassy employees in Tokyo instead of exhibiting solidarity with the Japanese nation.

I didn't remain silent in the face of this accusation, immediately retorting that I was certain that for each Israeli embassy employee evacuated out of Japan, at least two Israeli physicians and rescue workers were generously flown into Japan as part of an emergency aid mission. I continued and stressed that there are very few countries in the world that are more willing to send emergency rescue teams anywhere in the world on short notice than Israel. This went on for several minutes, each of us passionately defending his own country until we both came to our senses and burst out laughing.

Group identification can be a temporary phenomenon. People do move from one job to another, one city to another, and sometimes they immigrate to another country altogether. But collective emotions for previous collectives can often persist even after we are no longer part of the group with which we are identifying. This is because the advantages that groups and the individuals of which they are composed obtain from collective rational emotions are to a certain extent magnified if they are unconditional and unchangeable. Without those characteristics the threat that the collective can project against rival groups will be far less effective. Think for example about immigrants who stubbornly cling to their previous national identifications. This phenomenon is exhibited even more dramatically if the group to which we currently belong is in conflict with our previous group.

This subject brings to mind a very perturbing story I heard from my father, who worked as a bank clerk during the Second World War. One of the bank's customers, a Jewish refugee from Nazi Germany, punctually arrived at the bank once a month on the same day and at the same time to transfer one-third of his monthly salary to a mysterious address in Germany. One day my father asked him about the purpose of these money transfers. The bank customer stretched himself up as straight as he could, like a soldier standing "at attention," and proudly proclaimed: "I may not have the right to participate directly in the war for the defense of the German homeland, but I still regard myself as having an obligation at least to support the German war effort financially."

Particularly strong evidence for the evolutionary roots of collective emotions is the fact that we seek to create collective emotions again and again, even when identification with a particular group does not serve any vital interest. An example of this is fan identification with a sports club.

The sports entertainment business is mainly focused on creating collective emotions that function as anchors. Sports fan clubs serve no real goals (in contrast to labor unions, for example, which protect their members from exploitation by employers; or nation-states, which protect their citizens from external threats). The common goal of sports fans is entirely virtual-victory on the part of the team that they support. But sports teams are not really ends in themselves. They exist to create collective emotions in society. The depth and strength of such emotions can only be appreciated by going into a throng of fans in a packed stadium, fans who are on their feet and roaring at the top of their lungs celebrating a goal that has just been scored.

To what extent do these emotions influence the games themselves? An interesting study published in 2005 by researchers at the University of Chicago and Brown University looked into the decisions taken by referees in a large sample of soccer games. The study focused particularly on time extensions at the end of regulation time, decisions that are solely in the purview of the referees and for which there are no definite regulations in the rules books.

The researchers discovered that referees tended to give time extensions favoring the home side. Teams with a lead late in a game prefer brief extra time periods, while teams that are behind want longer time extensions (the more time they are given, the greater the chances that they can score). According to the study's findings, when the home team had a late lead, extra time periods were short, but if the visitors held the lead, the referees exhibited much more generosity in adding extra time. Since fans of the home team tend to outnumber fans of the visiting team in most stadiums, it is reasonable to postulate that this home team bias on the part of referees is due to the powerful collective emotions "radiated" by the home team fans to the referees.

Why do women exhibit relatively little interest in following sports teams? As noted earlier in this chapter, the primal source of human collective emotions comes from a need for the assistance of a group in obtaining vital resources, especially in the context of a group hunt. Since hunting is primarily a male pursuit, the need for collective emotions is often greater in men than in women. This might explain why men tend to follow sports more than women, and why men tend to be more nationalistic than women.

The collective emotions we have dealt with so far include rage, empathy, and collective admiration, but there is at least one more collective emotion: insult. Collective insult can sting more than personal insult.

Imagine, for example, applying for a job and receiving the following response:

Dear Mr. John Doe

Thank you very much for your interest in our company.

Unfortunately we cannot offer you a job because, frankly, you have low scores on standardized tests. Whether for permanent or temporary employment, our company has a policy of hiring only candidates with higher scores.

We wish you the best of luck in your pursuit of a job commensurate with your skills.

This is a painful, embarrassing, even outrageous letter. But imagine getting this letter instead:

Dear Mr. John Doe

Thank you very much for your interest in our company.

Unfortunately we cannot offer you a job because you are black. Whether for permanent or temporary employment, our company has a policy of not hiring African-Americans.

We wish you the best of luck in your pursuit of a job commensurate with your skills.

Most of us would regard this letter as much more outrageous than the previous letter, even though both are highly insulting. Both letters say more about the company than the applicant, but the second letter says nothing about the personal qualities of the applicant, rejecting him solely on the basis of collective identification. Why do we regard the latter as more insulting? Why would a black person feel more offended by the second letter? One possible answer is that the first letter contains what could be considered a rational reason for rejecting the applicant, while the second is bereft of any rational justification. But this is insufficient alone as an explanation. Imagine that the second letter included the following text:

> *In the past we have noted that African American employees steal 20 percent more office equipment than white employees.*

This seemingly offers a rational justification for rejecting the applicant, but it is just as outrageous, if not more so, than the letter without this addition. That is because rejecting an applicant based on his race engenders collective insult. It insults not only the individual, but his collective identification. This is another example of the way that collective emotions can be more powerful than individual emotions.

Are collective emotions rational at the individual level? Very often they aren't. Uncle Walter didn't gain much from his temporary enthusiasm at the Nazi rally, and a rational self-interested employee would care more about personal criticism than about criticism based on racism. But collective emotions are rational in a different sense. They are collectively rational. When viewing the group as a unit, this unit will do better when its elements experience collective emotions. Evolutionary forces operating on groups (rather than genes) can shape collective emotions. We will be discussing this form of evolution in our next chapter.

11

THE HANDICAP PRINCIPLE, THE TEN COMMANDMENTS, AND OTHER MECHANISMS FOR ENSURING COLLECTIVE SURVIVAL

THE MOST WIDELY ACCEPTED MODEL OF EVOLUTION HAS AT ITS CORE two main elements: mutation and selection. Mutation causes random changes in the characteristics of organisms from one generation to another. Selection is the mechanism by which "good" mutations spread in a population while "bad" mutations slowly die out. Individuals who possess good characteristics are better at surviving than those who lack them, ensuring that they have more offspring.

We usually think of evolutionary forces as shaping the characteristics of individuals (or genes), but mutation and selection also influence the evolution of societies. Societies with positive characteristics (such as social structures and values that preserve cohesion) will survive better than societies that lack them. The latter will be defeated in battle more often and will be abandoned more often by individuals.

Researchers in biology and social science are increasingly using models of group evolution to understand social structures among animals and humans. The two main evolutionary models that have been developed in this field are called group selection and kin selection. These models differ in much more than name alone: using one instead of the other can sometimes lead to completely opposite conclusions.

Take, for example, the question of whether or not humans may one day attain a life expectancy of 1,000 years. According to the kin selection model, this is eminently possible. Mutation will bring about random genetic changes that over time will grant humans immunity to virtually every known disease. Those lacking the life-extending genetic mutation will die out, leaving only individuals with a 1,000-year life span.

From the perspective of the group selection model this sort of development is inconceivable. Societies composed of people with 1,000-year life spans will fail to benefit from generational change and become developmentally "frozen."

They will suffer from perpetual resource constraints as their population levels skyrocket, leading to wars that will kill off many more people than in societies like ours, in which life expectancy is only 80 years.

Researchers have heatedly debated the legitimacy of the group selection model. Critics claim that thinking of societies or groups as autonomous individuals is fundamentally wrong. In this view, only individual animals (or human beings), along with their genetic compositions, can be conceived of as individuals subject to evolutionary pressures.

I find this approach too rigid. The "what counts as an individual" question is a philosophical one that has no single unequivocal answer. Consider examples such as ant colonies or corals, where what should be regarded as the "individual" level is far from immediately clear. In many cases an entire ant colony can be studied as an individual more usefully than if it is regarded as a collective formed from individual ants. The same reasoning applies to corals as a whole versus their constituent polyps.

In fact, one can think of a single human being as a colony composed of the individual cells in his or her body. This approach is increasingly being adopted in medical studies. There are articles in scientific journals analyzing competition taking place between cells in living creatures using game theoretical models that have successfully explained various pathological phenomena, including cancerous growths.

The subject of altruism is one of the more perturbing questions being studied by social scientists (including economists) and biologists. The kin selection model can explain why an individual might sacrifice himself for the sake of a family relation (such as a sibling or an offspring). Altruism of this sort can survive and spread in a population because the survival of family relations is in effect the survival of genes, given that family members share a similar genetic pool. The survival of genes is equivalent to the survival of behavioral characteristics.

But how can true altruistic phenomena, in which individuals help other individuals with whom they share no genetic heritage, be explained? The moral principal of aiding others (even if they are not family members) is nearly universal. It can be found in all cultures and religions. What advantage is there to individuals in wanting to help others and exhibiting broad solidarity? The psychological reward from the satisfaction of helping others cannot be sufficient to explain this. This positive sensation is a symptom of the fact that giving contributes to individual survival, just as the pleasure we take from eating sweet chocolate is a symptom of the fact that sugar (in proper moderation) is necessary for our survival. But in both cases the sense of satisfaction is not an explanation.

The desire to help others (even if they are not family relations) exists in other animals, not just in humans. The story of the Battle at Kruger described in the previous chapter is an excellent real-life example. The Arabian babbler, a bird native to desert areas in the Middle East, provides us with another. Babbler flocks have very complex social structures. They live in "communes," with collective sleeping areas in which the adults share in the burden of raising all the chicks of the flock. They assist each other in incubating eggs, seeking food for the chicks, and providing for the common defense of the entire flock's brood of chicks. Each adult bird is, in effect, investing heavily in providing for the flock at the expense of providing for his or her direct offspring. Could all this possibly exist simply because of the psychological boost that babblers get from helping others?

The kin selection model and the group selection model can both provide evolutionary explanations for the emergence of the behavioral trait of helping others. Both humans and babblers receive a personal reward in the form of increased chances of survival as a result of helping others. Altruism in societies promotes reciprocity. There is no room for freeloaders in societies that exhibit a preference for reciprocity (in addition to a desire to help others). Individuals who lack the desire to help others are punished by social ostracism that harms their prospects for survival. In contrast, those who give are usually rewarded by receiving the support of others.

Research conducted with the use of fMRI shows that the brain reacts to social ostracism in the same locations and to the same extent that it reacts to disease and the threat of severe danger. In other words, social ostracism and existential threats lead to the same distress reaction.

Of course, there are also differences between human societies and babbler flocks. A babbler bird who invests all his energy solely in providing for his chicks will not be helped by the other birds, and risks being physically chased out of the flock. The communal living structure of the babblers and the close cooperation that takes place between them enable close and efficient monitoring of the behavior of each individual.

Human societies, on the other hand, are less collective and much more individualistic. Thus it is difficult for humans to monitor the altruistic behavior of the peers, whereas babblers do so relatively easily and comprehensively. This can weaken the advantage of altruistic giving among humans.

There have been attempts to create human societies that are communal to the same extent as those of the babblers. Such communes proliferated in the United States in the "hippie" culture of the 1960s. Kibbutz collectives in Israel included communal sleeping quarters for children well into the 1990s. The general record of failure of human communes to sustain themselves over time indicates that babbler-style living arrangements are not natural for humans.

Another example of altruism among birds can be found in the behavior of starlings. In contrast to the babblers, starlings jealously protect their mates and offspring. They do not care for the chicks of other starlings, and they react aggressively to any attempt by rivals to steal their mates. But starlings behave with impressive bravery when it comes to external threats. If a predator approaches a flock of starlings, the first bird who notices this will emit loud cries in order to warn the other members of the flock. Doing so is not only a waste of energy from the selfish perspective of the individual who has volunteered to warn the others, it actually increases his personal danger by drawing the attention of the predator.

Zoologists place this sort of altruism in a different category from the altruism exhibited by babblers. The behavior of the starlings is related to the "handicap principle," first proposed by the biologist Amotz Zahavi.[1] The handicap principle posits that animals (especially males) will handicap themselves or place themselves in apparently dangerous situations to signal potential mates that they have genetic advantages, thus improving their prospects for mating successfully, outcompeting their rivals. Zahavi originally suggested the handicap principle as an explanation for the evolutionary development of the peacock's tail. Peacocks have extraordinarily impressive tails. Those tails, however, are also extremely heavy while providing in return no physical advantage in the peacock's natural habitat; in fact, the tails are so burdensome that they are disadvantageous.

The understanding that peacocks' tails are actually handicaps naturally led zoologists to ask why evolution hadn't long ago eliminated them. Zahavi's answer to this question was brilliant and original: the advantage of the peacock's tail, suggested Zahavi, is precisely that it is a handicap. Having such a long tail is not a luxury that every peacock can afford. Only the strongest, healthiest, and smartest peacocks can have efficient and easy mobility despite the limitations incurred by carrying a heavy tail. A large tail is in fact a signal indicating strength, health, and intelligence, attracting peahens seeking strong, healthy, and intelligent mates who will pass those genes on to their offspring and thus increase their prospects for survival. A peacock with a very long and heavy tail is rewarded with a steamy love life with several peahens while passing on genes that increase his chicks' chances of survival. The male offspring of such a peacock will, of course, also have long and heavy tails.

My colleague Yair Tauman has made use of the handicap principle in his research to explain the tendency of founders of high-tech start-ups to drop out of college before completing their degrees, even when they are very close to the end of their studies;[2] Microsoft founder Bill Gates and Facebook founder Mark Zuckerberg, both of whom dropped out of Harvard, are only two of the most prominent examples. In Tauman's model, such people, who are well aware of their own talents, find it advantageous to drop out because this "handicaps" them in a way that sends a positive signal to potential investors. In effect, they are saying that they believe in themselves and their ideas to such an extent that they are willing to forgo the job market advantage conferred by an academic degree.

The handicap principle also explains the altruism exhibited by starlings. Starlings do not strut around showing off useless tails, but they do show off by loudly helping their flocks avoid predators. The closer a starling allows himself to get to a predator and the louder he emits a warning call, the better off he is in signaling that he has advantageous genes, thus increasing his chances of impressing potential mates.

In this regard, humans are no different from starlings. A few years ago my nephew Ro'i volunteered to serve in an elite military unit with very selective entry requirements. He and his friends from the unit wanted to celebrate their completion of the unit's very stringent and demanding training course by holding a large party in a club in Tel Aviv, and they went from club to club to bargain the best possible price for reserving a place for a night of partying. They certainly managed to get a good deal—one of the largest and most luxurious clubs in the city offered not only to let them hold the party on its premises for free, it also gave each soldier in the unit a valuable gift. All it asked in exchange was for the soldiers to agree to let the public join in their party for a fee. The club made a great deal of money that evening from the entrance fees it collected. Hundreds of young women came to the party in the hopes of meeting a brave and strong young man in an elite military unit. The presence of so many women attracted an equal number of young men.

One might make the claim that both starlings and humans, such as soldiers volunteering for dangerous military units, do not exhibit true altruism because they know exactly what they will receive in return while true altruism involves giving without expectation of receiving anything in return. In fact there are biologists who claim that pure altruism does not exist and cannot possibly exist in nature, on the grounds that any behavior that confers no advantage to the individual performing it will eventually go extinct through natural selection. Obsessive altruists, individuals who give and only give (and there are people like that) cannot survive from the evolutionary perspective because they will come to the aid of others in dangerous circumstances but refuse to accept aid for themselves.

There is, however, an evolutionary explanation at the genetic level that can justify pure giving in social groups with very little genetic heterogeneity. In such groups, granting assistance to another member of the group is similar to helping a daughter or a brother. Helping another individual in a genetically homogenous population to survive, under this theory, is like helping yourself survive, because in effect you are helping your own genetic inheritance to survive and spread. Researchers debate whether this explanation is applicable to the sort of altruism exhibited by starlings, but it is widely accepted as explaining the behavior of social insects, such as ants and bees, who long ago lost the capacity for individual reproduction and instead loyally serve their queens. Altruism and giving, it should be noted, are more prevalent among humans in ethnically homogenous societies.

About a year ago I was invited to visit the University of Oslo in Norway. The Norwegian government had invested a large sum of money in a research study focused on a comprehensive comparison between the Scandinavian economic system and the more free-market oriented systems that are prevalent in other developed countries. This research effort was not entirely politically neutral. I got the impression that the Norwegian government was trying to justify the advantages of the egalitarian Scandinavian system to itself, its citizens, and the rest of the world.

In all honesty, anyone who has ever visited Norway or Sweden will find it very difficult to argue against the Scandinavian approach. The Scandinavian countries are blessed with strong economies, excellent health and education systems that are available for free to all citizens, and virtually no poverty or crime. Taxation levels in Scandinavia are among the highest in the world, yet tax evasion is completely negligible.

When asked for my opinion, I said that what needs to be studied is whether the extraordinary success of the Scandinavian system is a result of the system itself or of the public that chose this system. It would be very difficult, I claimed, to transplant the Scandinavian system to other countries. The Scandinavian nations are much more homogeneous than most Western nations, whether the homogeneity is measured ethnically or culturally. They developed historically from small Viking tribes with traditions of egalitarian sharing, tribes that over time grew to the size of nations.

Countries (such as the United States) that have to contend with much more ethnic and cultural heterogeneity will find it difficult to adopt a Scandinavian-style economic system because doing so would involve a significant amount of cross-ethnic and cross-cultural giving. Research recently conducted by the National Bureau of Economic Research in the United States studied patterns of community charitable giving in American neighborhoods. Ethnic diversity in a neighborhood was correlated with small amounts of community charitable giving. A 10 percent increase in ethnic diversity in a neighborhood is associated with an average drop of 14 percent in community charitable giving.[3]

The genetic-level evolutionary explanation of the spread of altruistic giving in populations is based on three elements. The first is deterrence: an individual lacking any sense of solidarity or desire to come to the aid of others will be ostracized from social interactions and thus pay a very high personal price for such behavior. In early human nomadic hunter-gatherer societies that price was equivalent to a death sentence. A successful hunt requires close cooperation among a band of hunters. A man in hunter-gatherer societies who failed to cooperate on hunts or refused to share with others would very quickly find himself starving to death with little prospects of reproducing. Those behavioral traits would thus die out.

The second element is the handicap principle. The very act of conspicuous giving increases an individual's chances of reproducing. The third element is the fact that in genetically homogeneous environments, giving to others serves the interests of propagating the altruist's genes.

The group selection model provides us with a smooth, direct, and simple explanation for the evolutionary survival of altruistic behavior. This model posits that mutation and selection operate at the group level, as opposed to the individual (or the genetic level). Groups that fail to place moral value on mutual assistance will go extinct faster than competing groups.

Imagine a battle between two tribes for control of vital natural resources. One tribe maintains a strong moral imperative for mutual aid among its members, while the other tribe believes that each individual should look out only for himself. It is not difficult to predict the outcome of the battle. Keep in mind, however, that even at the group level there has to be some amount of moderation for the principle of altruism to be beneficial. A tribe whose moral code calls on each individual to sacrifice him or herself for the sake of others in every possible situation will go extinct faster than a competing tribe embracing a less sweeping code of altruism.

This is one reason that religion has been such a powerful force in human history: it creates a social cohesion that benefits the col-

lective of its adherents. The Ten Commandments are a fine illus-
tration of this principle, having helped ensure the survival of the
relatively small world population of Jews, and later the survival of
much larger communities of Christians and Muslims. Not coincidentally, almost all of the content of the Ten Commandments have
been adopted widely throughout the world in the form of religious
or social precepts.

At their heart, the Ten Commandments operate based on three
mechanisms: (1) ensuring the physical existence of the group, along
with its social cohesion; (2) incentivizing reproduction; and (3) providing disincentives to leaving the group.

The first three commandments are there to ensure the primacy of
this ethical code above all others. Of course, populations that take
their code more seriously are more likely to follow it, and thus to survive. The next seven create a social contract, enforcing prohibitions
on theft, adultery, and murder, as well as creating mutually beneficial
relationships between family members and neighbors.

The importance of many of these commandments to the well-
being of a community is self-evident. But there are a few that bear
further analysis. The fourth commandment, "Remember the Sabbath
day to keep it holy," has an important role in preserving the group.
The Sabbath day is a day of rest during which an individual's main
attention is focused not on him or herself but on the collective; it thus
preserves group cohesion. The commandment explicitly relates to the
"other": "your son, your daughter, your servant, your livestock, and
the sojourner who is within your gates." This commandment also
encourages economic relations within the group, reducing the danger
of individuals abandoning it to join other groups. An employee from
within the group will seek an employer from the group, who observes
the Sabbath rest and grants the employee the same day of rest. That
employee will similarly find it difficult to work for someone who is
not a Sabbath observant and therefore may expect him to work on
the Sabbath. This creates mutual dependencies within the community, reducing the chances that people will leave the group.

The fifth commandment, "Honor your father and your mother," stands out among these commandments. It is the only one that promises a reward, "that your days may be long," to those who follow it. This creates a very clever social mechanism—an intergenerational contract—that functions as a strong incentive to have children.

At first glance it may be unclear how honoring one's father and mother serves the goal of group survival, and why such a tempting reward as long life should be offered for those who keep this commandment. Elderly mothers and fathers require honor and assistance long after their fertility has waned. From a purely evolutionary perspective you should perhaps invest time and effort exclusively in your children, not your parents, for the sake of both your genetic survival and the survival of the group. You might even be tempted to think that ignoring your parents and leaving them to fend on their own would be more beneficial for the group than helping them, because elderly parents consume precious and limited resources while contributing almost nothing to the group in return. They cannot protect the existing generation, and they do not create new generations.

But this line of thinking turns out to be wrong. An ethical code permitting hostility or even indifference to elderly parents creates a disincentive to having children, threatening the group with extinction. The true meaning of the reward "that your days may be long" then becomes almost self-evident: honor your father and your mother so that your children will in turn honor you when you become elderly. This intergenerational contract is amazingly similar to a pension scheme. When we urge our adult children to keep visiting their grandparents and inquire about their well-being, we are, even if subconsciously, reminding them that the intergenerational contract also applies to us. In summary, the fifth commandment is not only a way of ensuring that parents are cared for, it also provides an incentive to have children, without which the group cannot continue to survive.

The Bible and Talmud contain many more rules and regulations, over and above the Ten Commandments, intended to preserve group cohesion and stability. These were especially necessary for a dispersed people such as the Jews, who lacked a national territory of their own and faced many incentives to leave the group and join instead the broader cultures in which they were located.

The rules dictating kosher food are an interesting example of this. Many believe that they were intended simply to protect people from unhygienic food sources, but their true aim was preserving group cohesion. In virtually every culture and every era, dining was considered an important social activity that was conducted in a group setting. The kosher food rules greatly limit opportunities of Jews and non-Jews dining together, thus limiting social interactions between Jews and non-Jews in general. This by itself reduces substantially the chances of coming into contact with incentives to leave the group. Completely forbidding common dining between Jews and non-Jews might have created unwanted antagonisms. The construction of a complex set of seemingly arbitrary rules limiting which foods may be eaten succeeds in limiting contact in a more subtle way.

The above analysis is based on the group selection model of evolution. But every evolutionary model (including those operating at the level of groups rather than individuals) requires mutation in addition to selection. In the group selection model, the role of mutations is in ensuring that the group does not remain forever static in its norms and behaviors.

This is especially important under changing environments. Liberal societies that are tolerant of minorities, public protests, eccentric behavior, and free expression of ideas, enable mutations to contribute positively. They facilitate the adaptation of the group to its changing environment. Many of the most important social changes in human history started out as anomalous behavior relative to social norms. Fundamentalist societies that aggressively crush every attempt to introduce change block social mutations from taking effect. They lose their ability to adapt their norms and values to changing environments, greatly reducing their chances of surviving social genetic competition.

12

KNOWING HOW TO GIVE, KNOWING HOW TO RECEIVE

The Full Half of the Cholent

IN RECENT YEARS SEVERAL STUDIES HAVE BEEN CONDUCTED THAT attempt to increase our understanding of the mental processes we undergo when we give altruistically without receiving any compensation in return. One of the most interesting studies in this field was conducted by two friends of mine, Uri Gneezy and Aldo Rustichini.[1]

Gneezy and Rustichini set out to test the commonly accepted economic assumption that material incentives always increase people's motivations to undertake tasks; a slightly weaker form of this assumption postulates that incentives can never reduce motivations. To do this they conducted a field experiment. Field experiments differ from laboratory experiments in that they are conducted in the normal environments in which subjects interact in their daily lives. In many cases, the subjects are not even told that they are participating in an experiment. The advantage of field experiments is that their results are generally considered more conclusive than laboratory experiments. On the other hand, in field experiments the experimenter usually has much less control over the environment in which the subjects are located than in laboratory experiments. In many cases the experimenter may know nothing about some of the individuals related to that environment. I'll present two such experiments conducted by Gneezy and Rustichini.

In one experiment, Gneezy and Rustichini followed the activities of a group of children who, as part of a compulsory high school project, went from house to house to collect donations for disadvantaged youths. The children were divided into two groups. Group A was the control group, in which the children were told, as is normally done, that the donations they collected would all go to a central charity fund to be disbursed to the needy. The children in Group B were told that they would each be paid 20 percent of the donations they succeeded in raising, in compensation for the time and effort they were putting in. Both groups set out simultaneously and returned at the same time.

The outcome was surprising but also logical and reasonable. The children who were offered compensation for their efforts ended up on average raising significantly less than the children in the control group. The monetary incentive lowered motivation, in contrast to the usual assumption that monetary incentives can only increase task motivation.

Many of us would probably have guessed correctly that this would be the outcome of the experiment. It tells us less about the actions of the children than about our intuitions regarding the relationship between mental compensation and material compensation. Once the children in Group B were told that they would be paid for their efforts, the mental compensation and satisfaction they received from the good deed of being involved in giving to the needy was irrevocably reduced. Instead of a volunteer effort for a cause, the task became a salaried job. And as a job, the pay was rather lousy for the amount of effort required. If it were offered to them from the start as a job, the children would probably have turned down both the work and the salary. Since they had no choice but to undertake it, they put in little effort with correspondingly dismal results.

My friend Dan Ariely once used the following analogy to describe the situation in which the children in Group B found themselves: at the end of a splendidly enjoyable evening at home with a pair of friends who were invited to dinner, just before the guests leave after the accepted goodnight handshakes and hugs, the wife hints something to her husband, who takes out his wallet, turns to you and asks: "I nearly forgot, how much do we owe you for the excellent dinner?"

Just as monetary incentives can reduce mental motivations for altruistic behavior, monetary fines imposed for selfish behavior can actually reduce the mental fine we would otherwise feel and thus induce us to act more selfishly. That was the subject of Gneezy and Rustichini's second experiment, in which day care centers in the city of Haifa were asked to track the number of times per month that parents arrived late to pick up their children at the end of the day.

With that baseline data, the experimenters suggested imposing fines on late parents over a period of a month, to see how the fines would affect their behavior. The number of late arrivals, and their durations, of each parent over the month were totaled up to determine the fine to be charged, at a reasonable rate. The results of this second experiment were consistent with those of the first experiment. Imposing fines, rather than reducing lateness, actually increased it. Since they were now paying for the amount of time during which the children were in the day care center past closing time, parents regarded that time as akin to a "paid babysitting service." That relieved them of the discomfort and shame they had previously felt when arriving late.

The insights gained from Gneezy and Rustichini's experiments are very important for understanding the behaviors of organizations and private corporations. We only rarely use them, however, to create efficient incentives. At the individual level, in relationships between friends, there is usually a large emphasis on keeping track of favors (monetary or otherwise), with each individual striving to repay favors as soon as possible. In most cases this is not due to a pure-hearted motivation to be a giving person. Quite the opposite. In fact it is a selfish trait. Because favors are socially regarded as acts that should be mutually repaid, the benefactors of favors seek to reduce the "favor debts" that accrue as quickly as possible, even to the point of ruining the satisfaction the doer of the favor gets from the act of giving. When the needs of givers and receivers are consciously taken into account, the giving and accepting of favors between relatives and friends usually become much smoother and more stable.

When I was a child, my family would regularly have meals on holidays at my grandmother's home in Jerusalem. My mother's seven brothers and sisters, along with their families, would also attend. The central dish was always the traditional Jewish slowly simmered stew called cholent. Each of the sisters, including my mother, would pre-prepare a separate ingredient of the cholent. All the ingredients would then be placed into an immense pot, cooked, and served to the forty-plus guests in my grandmother's one bedroom apartment.

Each sister habitually prepared twice as much as necessary. After we had stuffed ourselves to complete satiation, with the cholent only half-eaten, an argument would begin over how to divide the remaining half. Initially, each sister would go to great lengths to explain why she couldn't possibly take home even a single spoonful: she's on a diet, there is no room at all in the fridge at home, and so on. The most commonly repeated sentence at this stage was "with me it will end up in the garbage."

Serious negotiations would begin in the second stage. "Matilda, that really isn't nice of you. I took home all the leftovers last time. If you don't take what I prepared this time, I won't talk to you." In the third stage, compromise would finally be attained: "All right, I'll take some of this here if you take the rice and beans over there."

We all loved the cholent—every bit of it and every ingredient that went into it. But the pleasure in eating the cholent was insignificant compared to the satisfaction we took from giving. We needed to give to such an extent that it was worth arguing over it. Sometimes arguments would last weeks, with each woman mentally remembering exactly which sister refused to take a particular leftover item a year earlier.

One time, the leftover division ceremony started off with the usual round of exclamations of "I simply can't take home anything!" But Aunt Rachel uncharacteristically did not join in. Aunt Matilda, who noticed this immediately, was quick to try to make use of the unexpected opening. "Here, you really must have some of this," she said to Rachel while pushing two overflowing bags of leftovers into her hands. Rachel grasped the bags and simply said, "Excellent, thank you very much."

A stunned silence descended on the room. We all looked at Aunt Rachel as if she had possibly lost her mind. Aunt Dina worriedly inched closer to my mother and whispered in her ear that perhaps Uncle Moshe (Aunt Rachel's husband) was experiencing financial difficulties in his carpentry shop. Their entire household must be scrimping every penny!

After Moshe had assured everyone in attendance that his carpentry shop was doing better than ever and that the household finances were as solid as a rock, thank you very much, it slowly dawned on the sisters that it was Rachel doing Matilda a favor by agreeing to take the leftovers home, not the other way round.

This changed everything. The family cholent gatherings became much calmer. Half of the food served still had to be divided at the end of every meal, but those divisions became much more equitable. Each sister was happy giving and also, sometimes, taking.

The point of the cholent story, of course, is that if giving is its own reward, then sometimes taking can be a favor. If my aunts had continued to negotiate purely in economic terms, each one should have been happy to take as much of the stew home as possible, and perhaps bickered over the last drops of it. By taking a sensitive, emotional approach, Aunt Rachel had short-circuited the whole problem, leading instead to a resolution that benefited everyone.

PART III

On Love and Sexuality

13

THE SPRAY THAT WILL GIVE US LOVE

On the Hormone That Creates Trust
and Neutralizes Suspicion

OXYTOCIN IS A HORMONE RELEASED BY BREASTFEEDING WOMEN AND THE infants whom they are nursing. Studies conducted on primates reveal that it is responsible for the bond formed between mother and child immediately after birth, before they have managed to forge a deeper connection. The hormone is also released by both sexes during sexual orgasm and is therefore often called the "love hormone." Oxytocin is a wonderful evolutionary mechanism that increases the chances that a newborn infant will survive, thus passing on genes from one generation to the next.

Before we have children of our own, many of us marvel at the ability of new mothers to find the energy resources needed to care for a newborn infant after nine exhausting months of pregnancy. They manage to do this within seconds of what is often a difficult and draining childbirth experience, without having had any opportunity to form an emotional bond with their child.

This is accomplished because the evolutionary development of primates, including humans, has supplied females with a hormone that makes bonding between mother and child completely instinctive. It even enables an infant to understand the importance of finding his mother's breasts minutes after emerging into the world; infants are born with the instinct to suckle their mother's milk.

Oxytocin is also connected to two known developmental disorders. An imbalance of oxytocin, especially a deficit of the hormone in the brain, is characteristically identified in those suffering from autism spectrum disorder. A lack of oxytocin is one of the reasons that children with autism spectrum disorder experience difficulty in exhibiting empathy toward others, understanding social situations, and trusting those who are close to them.

The opposite condition is noted in individuals suffering from an extremely rare neurological condition called William's syndrome, which is characterized by a range of physiological and mental disturbances. These include heart conditions, digestive tract disorders, and elevated blood pressure. Their IQ levels are typically limited to 60 to 90 points, but their social skills are impressive. They exhibit empathy and the ability to recognize emotions in others at levels that are far superior to those of normal humans. They are also willing to trust others, even total strangers, almost blindly. Children suffering from William's syndrome express love to everyone around them. This makes them vulnerable to sexual exploitation, as the exaggerated trust and desire to please others that they feel makes them easy prey for pedophiles. Neurologists posit that elevated production of oxytocin may be partially responsible for the social behaviors of individuals with William's syndrome.

Given the important role oxytocin plays in creating bonds between mothers and infants, as well as its relation to social development disorders, it is reasonable to suppose that it also influences social behavior in healthy adults.

Oxytocin is a benign hormone that is harmless when introduced into the body in small doses (this is usually accomplished by using nose drops, similar to the nose drops used to ease the symptoms of the common cold). The Zurich experimenters had two groups of subjects play the trust game: one group received a dose of oxytocin before playing the game, while a control group was administered a placebo containing all the same elements except the active element. The results were unequivocal: members of the group receiving doses of oxytocin achieved much great cooperation. This cooperation was exhibited in both directions: proposers offered more (they trusted their counterparts more relative to the control group) and receivers gave proposers a larger share of what they received in return (they were more generous).

To rule out the possibility that oxytocin was simply relaxing the subjects in the experiment and thus indirectly making them more amenable to cooperation, the experimenters repeated the experiment using wine instead of oxytocin to relax subjects. The wine did indeed have the effect of making subjects feel more relaxed, but it had no effect on the amount of trust or generosity they exhibited.

Oxytocin, for all its wonders, can also have negative effects. An experiment that I recently conducted along with two of my students, Einav Hart and Shlomo Yisrael, revealed that oxytocin can reduce our abilities to recognize the intentions of others.[1] In our experiment we made use of the television game *Split or Steal,* also discussed in Chapters 3 and 4. The subjects in the experiment viewed video clips of the game. During the course of the experiment they were asked to guess which actions the participants of the game would choose on the basis of the brief dialogue they conducted prior to choosing either split or steal. One group of subjects was administered oxytocin while a control group received a placebo. Although subjects were unable to tell whether they had received oxytocin or a placebo, those who were administered oxytocin were much worse at guessing the actions chosen by the *Split or Steal* participants than those in the control group. When we compared the reaction times of the two groups, we discovered that subjects receiving oxytocin invested much less effort in this task than those in the control group, making their guesses hastily. Why oxytocin should have this effect is fairly clear. We are most invested in identifying the intentions of others when we are suspicious of those around us. Since oxytocin dulls suspicions and boosts trust, it makes us more vulnerable to being manipulated by others.

The effects of oxytocin, both good and bad, make the use of the hormone potentially dangerous as a tool for manipulation. A spray called Liquid Trust, whose active ingredient is oxytocin, is now commercially available. It is promoted as a chemical that can influence buyers' decisions in market interactions. The Liquid Trust Web page describes it as ideal for salespeople, lonely individuals seeking love, and managers and employees who want to influence their work environments or seek rapid promotion. The product's advertisements promise "the world at your fingertips." Is there a practical way to outlaw the use of oxytocin? It is unclear how a law banning its use could be effectively enforced, given that the hormone has no taste or smell and is virtually undetectable when sprayed directly into the air.

The optimistic way to view the results of experiments on oxytocin conducted to date is to note that it can increase the chances of cooperation between people and in that way improve many economic and social interactions. But there is a thorn in this rose. Imagine negotiators representing two nations locked in negotiations on a contentious political dispute deciding to use oxytocin (of their own volition, not due to external pressure) in an attempt to improve the negotiating atmosphere and increase trust. Let's further imagine that the negotiations do indeed lead to a successful and satisfactory agreement due in part (but not solely) to the use of oxytocin. Would the public accept the legitimacy of such an agreement? I doubt it. Opponents of the agreement from both sides would claim, with some justification, that the negotiators were drugged into making concessions they would never have considered had they been fully sober.

Regardless of such imaginative scenarios, oxytocin illustrates the explicit connection between the way one feels and the way one thinks. It is a reminder that the hormone balance in one's body even influences careful cognition; thus all thinking is, at some level, emotional.

14

ON MEN, WOMEN, AND EVOLUTION

Testing the Myths

LOVE AND SEXUALITY ARE FAR AND AWAY THE MOST IMPORTANT emotional phenomena for our direct genetic survival. It is no surprise that nearly 80 percent of people surveyed by Daniel Kahneman and his colleagues in the course of their research on happiness reported that sexuality and love are the most decisive factors in their lives for achieving happiness.[1] The other rational emotions discussed in this book are important for evolutionary survival because they increase our fitness to our environment and our personal chances for survival. But love and sexuality directly contribute to our genetic survival by enabling us to reproduce and raise offspring.

Love is not a mechanism that is needed for reproduction in most animals, for whom sexual relations alone suffice. These typically involve brief sexual encounters, often only once with each mate, with males taking on little or no responsibility for caring for their offspring.

Many of us may also know humans who fit this description in their attitudes toward sexual relations. But most of humanity exhibits a different pattern of sexual behavior. The institution of marriage, a nearly universal cultural phenomenon, is a strong expression of the more typical human attitude toward love and sexuality. This distinction between human sexuality and that of most animals is related to the fact that raising a human child is a very long and complex process requiring the involvement of more than one parent.

While humans wait patiently for up to a full year or more for infants to learn to walk after their births, newborn gazelles are up and walking within two days of their births. Equine mares watch their newborn foals take their first steps within half a day of birth.

The life expectancy of gazelles and horses is shorter than that of human beings, but it still ranges up to about thirty years. Raising a human child to the point of complete independence from adult care and supervision takes about 20 percent of modern human life expectancy. Until about two hundred years ago, it required as much as 30 percent of life expectancy. There are virtually no other animals that go through such a lengthy juvenile period relative to their life expectancies.

From the evolutionary perspective there is no point in having offspring unless those offspring in turn have offspring of their own. Only a child who has reached independent adulthood can contribute to the genetic survival of his or her parents. If childhood were sufficiently short relative to the life span of a single parent, and demanded relatively few resources, mothers could reasonably care for their offspring on their own. The longer childhood lasts and the more one needs to invest resources in raising a child, the more important it becomes for the father, who also benefits (genetically) from having offspring who successfully reach adulthood, to share in the burden of raising the child.

Previous chapters looked into the roles that social emotions play in creating commitment. Anger, for example, helps us create credible threats. Love, in contrast, creates credible commitment for altruistic behavior toward mates, a commitment that is a precondition for parental cooperation in caring for offspring. From the male perspective, the commitment arising from love within a couple increases the chances that the child he is helping to raise is indeed his child, carrying genes that are similar to his, and not the child of another man with whom his spouse has had relations. Love and social structures that are built on stable monogamous relationships are the result of the large amount of parental energy humans need to invest for the successful survival of their offspring.

Human parents generally care simultaneously for children who were born in different pregnancies. This is not a phenomenon that characterizes other animals, whose offspring leave their parental nests before their mothers reproduce again. My colleague Motty Perry co-authored an excellent paper that used models of game theory to show that this phenomenon is responsible for the familiar structure of the human family, in addition to the commitments that members of couples exhibit toward each other.[2] Without these commitments, men would never know if the food that they have worked so hard to obtain and give to their spouses will be passed on to feed their children, as opposed to the children of other men from previous pregnancies.

Human childhood is very lengthy because human children need to learn complex social skills, over and above the physical and cognitive growth that all animals undergo as juveniles. Very few animals form long-term stable couplings with a single mate (hamsters and foxes are two noteworthy exceptions). The vast majority of animals have what we humans might call far more "steamy sex lives," based on casual sexual encounters. The sole purpose of their sexual interactions is procreation. Sexuality in these species is based on intense and sometimes violent "sperm competition" between males, along with selective female receptivity to the mating efforts of the males, with only the males deemed most fit on the part of the females succeeding in mating.

The specific characteristics of sperm competition between males vary from one species to another, depending on evolutionary developments. Competition between drones (male bees), for example, comes down to a total of about ten minutes out of their very brief lives. When a virgin queen bee is ready to mate, she enters a vigorous dancing state, drawing a swarm of drones. Only the strongest and quickest drones can succeed in mounting the larger queen bee and inserting their sperm into her. The drones die shortly afterward, while the queen bee stores their sperm for the rest of her life (up to thirty years) for use in fertilizing the millions of ova she produces.

Sperm competition between male mice is no less interesting. Its main expression comes after the act of mating has been completed. After inserting his sperm into a receptive female, the male secretes a sticky substance that essentially blocks the female's reproductive tract to prevent other males from successfully mating with her until his sperm has been fully absorbed inside the female. This strategy, reminiscent of the chastity belts that the knights of the Middle Ages once locked their wives in before going out to battle, increases the male's chances of successfully fertilizing a female with whom he mates and also incentivizes him to care for her offspring because he has greater certainty that her offspring are his.

Sperm competition strategies vary widely between species, but generally it is one of two kinds of evolutionary strategies for ensuring the survival of one's DNA. The other is a "marketing strategy" (think of the peacock's tail and other characteristics and behaviors that can be explained using the handicap principle) used to increase the attractiveness of individual males in the eyes of females.

Men and women have evolved differences in their emotional and sexual behavior due to physiological differences related to reproduction between the two sexes. Reproductive asymmetries between men and women are expressed in three main ways:

1. The maximal number of children that a woman can bear in a lifetime is well below one hundred. (The best documented historical record of the greatest number of children borne by one woman is held by a Russian peasant woman who lived in the eighteenth century and gave birth to sixty-four children through twenty-seven pregnancies.) In contrast, a man can theoretically father 100,000 children. Similarly,

while a woman can reach her maximal reproductive potential by mating with only one man throughout her life,
a man would need about a thousand women to attain his
maximal reproductive potential.

2. A woman knows with exact certainty who her biological
    children are: the children emerging from her womb. A
    man can never be certain whether the children borne by
    his spouse are indeed his biological children.

3. In the reproductive process itself mothers invest far more
    resources than fathers because mothers carry fetuses within
    them for nine months of pregnancy.

In addition to these three differences, men and women differ in
one more relevant physiological actor: men on average have greater
muscle mass than women.

To get an idea of the extent to which these physiological distinctions influence differences in emotional reactions and sexual behaviors
between men and women, I will review several widespread clichés,
taking a close look at each one. Keep in mind that the evolutionary
forces that have been shaping differences between the sexes long pre-
date the feminist revolution and the modern era. They existed before
human civilization arose, under conditions of a daily struggle for survival in which lack of close care for a child on the part of both parents
meant almost certain death for the child.

A brief discussion at the end of the chapter will also look into why
evolutionary gender differences in emotional and sexual behavior
stubbornly persist even in our modern world.

Cliché 1: Men are far more likely than women to agree to brief
one-time sexual encounters without emotional commitments.

The facts: A man can theoretically father a thousand times as many
children as any one woman can bear. In practice, men and women

have the same number of children on average for the simple reason that each child has precisely two biological parents. This brings about a situation in which men are in perpetual competition with other men in the race for greater fertility. From this perspective, a long-term commitment to one partner reduces a man's genetic survival potential because it limits the number of children he can have to the upper limit of children that his partner can bear for him. In contrast, women need only one man to attain their maximal fertility, and gain no advantage in having multiple sexual partners.

Cliché 2: Women have a greater need than men to express love.

The facts: As noted above, having sexual relations with multiple partners without any emotional commitment has no effect on the number of children a woman can bear. On the other hand, it does reduce her children's chances of survival, because if she has no partner with an emotional commitment to her and her children, then none of the fathers of her children is likely to contribute to the burden of raising the children. If she is alone in the task of providing for her children, they are likely to have less protection and less food than they would if they had a father helping to raise them. Procreation in general is more resource-demanding for women than men because a woman can have only one child every nine months, during which she needs to invest a great amount of energy in pregnancy and childbirth. As a result, women need to be much choosier than men in mating, and they need to ascertain that their mates will be committed to them and to their children.

Cliché 3: Women are more anxious than men when it comes to their health and the well-being of their children, while men become more nervous than women when their health shows signs of failing.

The facts: The image and stereotype of the "caring and worrying" mother is common in many cultures, and for good reason. Because women are more limited than men in the number of children they can have, they need to invest more resources than men in protecting the children that they already have. This is the evolutionary source of the "caring and worrying mother" figure. When all her children have achieved adulthood and her years of fertility are behind her, usually when she is in her fifties, a woman's task in directly ensuring her genetic survival is over. But a man at that age can still contribute to his genetic survival by fathering more children. Only death or disease can limit his further fertility. In other words, from the perspective of genetic survival, from age fifty and above only men have "something to lose," which may be the source of male hypochondria in their later years.

Cliché 4: Women are more jealous and suspicious than men of their partners.

The facts: It is nearly impossible to check this empirically. On the other hand, evolutionary explanations do not support the claim. Both sexes have good reasons to be jealous. A man needs to ensure that the children his partner bears, whom he is committed to supporting, are indeed his biological children. A woman needs to ensure that her partner will not leave her and commit himself to another woman in her place, leaving her children bereft of his protection and support.

But these evolutionary sources of jealousy differ between men and women, leading to differences in behavior. Several studies, including one by Monica T. Whitty and Laura Lee Quigley, have found that men are emotionally hurt most by sexual infidelity on the part of their partners, while women are more anxious to preserve emotional fidelity.[3] Interestingly, differences in emotional responses to infidelity between men and women are also expressed when they are the ones doing the cheating. Women who have intense emotional (but not sexual) relations with men who are not their partners feel stronger pangs of guilt than women who have extramarital sexual relations that do not involve emotional commitments. In contrast, men feel guiltier about sexual relations they have with women who are not their partners than about emotional relations. This can cause many couples to disagree about whether one of the partners has cheated, or whether jealousy is justified at all, even when they agree about the facts.

Cliché 5: Men are more likely than women to cheat on their partners.

The facts: An interesting research study conducted in the United States several years ago using DNA tests performed on newborn infants revealed that 5 to 10 percent of newborns are not the biological children of the men who are listed as their fathers.[4] Most of those men are completely ignorant of the fact that they are raising another man's biological child. This statistic, however, does not answer the question of whether more men than women cheat on their partners. The fact that men need more sexual partners than women to achieve their maximal fertility potential might lead men to be more receptive to opportunities for cheating, but that does not necessarily translate into more cheating in practice.

Imagine listing all the men in a particular town by their attractiveness to women, from the most attractive to the most slovenly and unappealing man you have ever seen. Although it is clearly not realistic, for the purposes of this thought experiment assume that all women would have the same preferences regarding the attractiveness of these men. Again, for the sake of the argument, suppose that in this virtual town each man is married to one woman and each woman is married to one man.

Now ask yourself which of these men has the best chance of conducting extramarital affairs with several women. The answer is obviously the men who are highest up in the attractiveness ranking. They can offer most of the women around them an opportunity for much "better" mating than the men to whom they are married. Women do not physically increase the number of children they can bear by increasing the number of sexual partners they have. What they do gain, however, is the opportunity to improve the genetic legacy they can give their children if they have relations with a more attractive man than their spouse. A man who is only slightly more attractive than her husband is unlikely to tempt a woman to cheat on her marriage, but George Clooney stands a good chance. Men, in contrast, can gain more by stressing quantity over quality, hence they will tend to be less choosy. It doesn't take a supermodel to tempt them to cheat.

What percentage of men, then, will manage to realize their dream of having an extramarital affair? The answer to that question depends on two variables. One is the distribution of "grades" that women give to the men around them for attractiveness, and the other is the extent to which women gain an advantage by remaining faithful to their husbands.

Suppose, for example, that the most attractive man in town is a perfect 10 in the rankings while all the other men are rated 5, and furthermore suppose that the advantage for remaining faithful to one's husband is low (which is the case in wealthy societies, in which women are not dependent on men contributing resources for the raising of their children). In this case the "adultery market" would be very simple. Nearly all the men (except for the top-ranked man) will be faithful to their wives while every woman except one will cheat on her husband (all of them with the same highly ranked man). In this case, despite the advantage that men clearly gain from having multiple partners, adultery would be mostly a female pursuit. This seemingly paradoxical situation arises from the market forces described in the example. All the men want to commit adultery but only one actually does so, while all but one of the women cheat on their husbands but only with the most attractive man in town.

This example is admittedly extreme, but it can be generalized. In any situation in which there is a small number of "stars" at the top of the attractiveness ranking who are far and away more preferred than their nearest competitors, there will be more women committing adultery than men. This may describe, to a certain extent, the true situation in wealthy and liberal modern societies with relatively weak economic anchors to maintain the traditional family structure. In traditional and religious societies individuals who cheat on their spouses pay a heavy price for their infidelity, with women usually punished more than men. The punishments can range from social ostracism all the way to execution. They significantly reduce the incentive for infidelity.

Cliché 6: Men are more competitive than women.

The facts: The Hebrew University of Jerusalem conducted a wide-ranging survey in 2003 to study the sex ratio of men to women at all ranks and levels, from students to full professors. The study produced an interesting set of data. A majority of students awarded the university's bachelor's degrees, 61 percent, were women. Among master's students women comprised an even larger majority, 62.5 percent. But the percentage of women obtaining PhD degrees placed them in the minority, at 46 percent. The representation of women among faculty members was even lower, 33 percent. Finally, the percentage of women who were full professors (the highest faculty rank at the university) was so low it was embarrassing—only 11 percent. These numbers did not surprise most people familiar with the composition of the faculty, but they did spark an intense discussion on the question of why the percentage of women drops so dramatically from one academic rank to the next.

A similar discussion at Harvard University a few years ago led to the firing of university president Larry Summers after remarks he made on the subject set off an uproar. Summers merely speculated that the lack of women in faculty positions in the sciences is related to differences in the competiveness exhibited by women and men. The Hebrew University discussion was less stormy. The data on the ratio of women to men completing bachelor's and master's degrees and the grades women were attaining in their coursework left no doubt that women are as intellectually capable as their male colleagues. Why, then, are women dropping out the higher up one looks on the academic ladder?

Some blamed the heavy burden that raising children places on women, a lack of day care opportunities for small children, and difficult hurdles that faculty members need to overcome for university promotion, which disadvantage mothers of newborns. Some accused the university of conscious or subconscious discrimination against women, claiming that men feel more comfortable in all-male working environments.

Pointing accusatory fingers at particular individuals or policies and blaming them for unbalanced sex ratios in corporations and institutions is convenient, but in my opinion this is an inefficient approach. It is convenient because it gives the mistaken impression that drastic changes can be immediately obtained if only aggressively enforced affirmative action policies are brought to bear. It is inefficient because it deals only with the supply side of senior jobs positions and not the demand side.

Several research studies conducted by behavioral economists in recent years have added to our insight in this subject. One such study, published by Uri Gneezy and Aldo Rustichini, revealed that men and women behave differently in competitive conditions.[5] The researchers gave men and women monetary rewards for solving maze problems on a computer. In the first stage of the study the participants received a set, uniform payment for every maze that was successfully solved. In this stage there were no sex-related differences evident—women and men were equally successful in solving the mazes.

In the second stage the offered payments terms were changed. Instead of a uniform payment for each successful solution of a maze, payments were based on the results of a competitive tournament. In other words, the participants were ranked in relation to others, with the payments they received depending on how high they were ranked. The money received by each participant now depended not only on his or her performance but also on the performance of others. In this stage men attained significantly better results than women. Not only that, women performed better in the noncompetitive stage of the study than in the competitive stage, managing to solve more mazes.

It is still unclear why the women performed worse in the competitive stage. One possible explanation is that they felt less motivated to make an effort to solve the mazes when the payments were based on tournament results. But another explanation is that the stress that was induced by the competitive environment of the second stage affected their abilities. Gneezy and Rustichini concluded that men perform better than women in competitive situations.

Another pair of researchers, Muriel Niederle from Stanford University and Lise Vesterlund from the University of Pittsburgh, also studied gender-based differences in competitive situations.[6] Participants in their study were paid to solve tasks requiring cognitive efforts—summing five two-digit numbers. This time, however, the participants had the option of receiving either a uniform payment based on their performance alone or a payment based on their performance in competition with others. A majority of male participants, 73 percent, chose the competitive payment method compared to only 35 percent of female participants who preferred that option. That large gap was independent of the relative performances of men and women in accomplishing the tasks in the experiment. Part of the gap stemmed from the simple fact that many of the female participants felt less comfortable being in a competitive situation, no matter how good they were at the task of summing five numbers. This is one of the most important points that emerged from the study: even women who were very good at the task and could have attained higher payments by choosing the competitive payment method preferred the noncompetitive method.

Several other studies, in addition to the two described in detail here, also indicate that men and women differ in their attitudes toward competition. There are also studies showing that women prefer avoiding negotiating situations much more than men.

Gender-based differences in attitudes toward competition may, if only partially, help explain the imbalance between men and women in senior jobs. Sherwin Rosen and Edward Lazear of the University of Chicago composed a very influential article in the 1980s comparing the promotion process in large organizations to sports tournaments.[7] An employee who wants a promotion in an organization needs to "defeat" several rivals in order to advance to the next level, just like a tennis player at Wimbledon. The higher one climbs in the hierarchy, the closer one gets to the spire of the pyramid. At each successive level the competition gets fiercer.

Rosen and Lazear give a very interesting explanation for the fact that the greatest leap in salary typically occurs between the penultimate level of the pyramid and its apex. At every other stage of the competition, they explain, if you get promoted, not only do you get a higher salary and more prestige, you also get another important prize, namely the right to compete for the next level in the hierarchy, where you will get even more money and prestige. If you get to the very top of the pyramid, you cannot receive this added prize, simply because it does not exist. There are no more levels to climb. The compensation for this comes in the form of a greater increase in salary in the move from second-in-command to the top position than the salary increases in all the other promotions. Otherwise, organizations would be reducing the incentive for promotion at the highest level of competition, hurting their chances of getting the best person for the top job.

Workplace promotion competition is usually not as transparent and blunt as in Rosen and Lazear's model. But it definitely exists, and the competition unquestionably gets tougher the higher up you climb in the hierarchy. That may be the reason that women, who on average avoid competitive environments more than men, often decide to bow out of the competition at a certain stage even when their talents and chances for promotion are equal to those of the men against whom they are competing. This is why gender-based affirmative action in general is unlikely to be the right policy to use for the goal of increasing the representation of women in senior positions in organizations and corporations.

In Rosen and Lazear's model, affirmative action is akin to lowering the bar by half a foot in a high-jump competition when the jumper is a woman. Doing so will not change the fact that there is a competition in the first place. It won't make women who prefer avoiding competition altogether feel any better about the process. In fact, it could have the opposite effect. Knowing that they are being judged by different criteria than those applied to men may harm their self-image and reduce the satisfaction they would otherwise get from winning the competition, reducing women's incentives to participate from the start.

A more efficient policy to adopt would be one that judges men and women using equal criteria but gives women a greater incentive to agreeing to compete in the first place. Possible incentives include giving women a "prize" for participating in the competition, even before the winner is announced, or offering a bigger prize for women who win the competition (which would translate into a higher salary or bonus given to women who attain promotions).

Sex-based differences in attitudes to competition undoubtedly developed during the course of evolution. Competitiveness gave males a greater survival advantage than it gave females. Competition between males for female mates is characteristic of many animals. Competitiveness gave human males an evolutionary advantage in genetic propagation. Acquiring food resources, hunting, and protecting families against predators and enemies are inherently masculine pursuits (given the more muscular frames men generally have in relation to women). They require a good deal of competitiveness. In a hostile environment, with scarce food resources that are difficult to obtain, a man who avoids competition risks death for himself and his family.

Cliché 7: Men are more likely than women to take risks.

The facts: Medical researchers studying the male hormone testosterone discovered an incredible relationship several years ago between the concentration of the hormone in the human body and the structure of the fingers of the hands. It is a very simple relationship that anyone can easily check by looking at his or her own hands. Place your right hand flat and spread open on a table top. Measure the length of your index finger, followed by the length of your ring finger, and calculate their ratio. In most men the index finger is shorter than the ring finger, giving a ratio of less than one. The smaller the ratio, the greater the concentration of testosterone in the body. This is a statistical relation that is not necessarily always true but it does occur in the vast majority of cases in a statistically significant manner.

High concentrations of testosterone are also statistically correlated with increased sex drive, stronger levels of concentration, and greater muscle mass. The hormone also has positive health effects, cutting down the concentration of lipids in the body and reducing heart attack risk.

On the other hand, testosterone is also linked to several negative phenomena, including many undesirable behavioral traits. People with elevated testosterone levels tend to be attracted to smoking and alcohol abuse. The chances that a man with high testosterone levels will develop a smoking habit are nearly twice as high as those of a man with relatively low testosterone. High-testosterone men also exhibit tendencies toward violent behavior and danger-seeking.

But that's not the end of the tale of the ring finger. Economists at the University of Cambridge compared the finger lengths of hundreds of financial "day traders."[8] Day traders, usually agents of investment houses and trust funds, buy and sell stocks at torrid paces. In many cases, using a method called "shaving" in the day-trading jargon, shares in a stock may be bought, held, and then sold in under a minute, sometimes within seconds.

Nearly all day traders are young men who work for only short periods of time for any one employer before being replaced. The Cambridge researchers tracked the work performances of several day traders and came to a startling conclusion: the lower the ratio of index finger to ring finger, the more likely a trader is to take risks in buying and selling stocks, and the higher the average profits he or she brings in. Even neophyte investors know that taking bigger risks can lead to higher average profits, but predicting the statistical likelihood that a trader will be willing to run great risks in the hope of netting large profits by looking at the lengths of fingers sounds completely ludicrous. Yet it is scientifically confirmed.

There are many additional research findings indicating that men and women have different attitudes to risk. An interesting series of studies trying to understand youth behavior were conducted in recent years. They focused, in particular, on the question of why obsessive thrill-seeking, provocative behavior, and thoughtless risk-taking are so prevalent among youths aged thirteen to twenty-three. Parents of children in that age range often find it difficult to understand their children's behaviors, forgetting that they themselves did the same when they were younger.

Research studies have shown that the brain of a youth over those ten years is still "a work in progress," during which new experiences, including extreme situations, are important for the development of an adult personality.

Significant differences were noted between the attitudes of male and female youths with regard to risk-taking. Young males take much greater risks than females in the same age range, and more risks than older men. This is a major reason that throughout history most of the blood spilled in battles has been the blood of young men.

The differences in risk attitudes between the sexes are also due to evolutionary developments related to competition between males over female mates. They can be explained by Zahavi's "handicap principle" mentioned in Chapter 11. Risk-taking by males broadcasts courage to females, a behavioral trait that indicates greater chances of success in protecting offspring from danger and obtaining food sources. This gives men who boast about their willingness to take risks in the presence of females an evolutionary advantage.

But the presence of another man also pushes men to greater risk-taking. Experiments studying the reactions of men in car race simulators indicate that the extent to which they are willing to take greater risks increases significantly when there is another man in the vicinity. Parents of teenage children who feel nervous every time their hormone-filled son takes the keys to the family car on weekends should feel calmer when their son is alone in the car than when it fills up with friends his age.

Natural selection has a role to play here as well. Displays of willingness to take risks in the presence of one or more other males are intended to intimidate potential rivals for mates. In the nineteenth century this trait became so pronounced that many young men lost their lives in armed duels fought over petty insults, all conducted with the approval of the establishment without anyone ever brought to trial.

Cliché 8: Men seek younger women, while women place less importance on the age of their mates.

The facts: Clear and incontrovertible statistical evidence shows that in most marriages husbands are older than their wives. But does this reflect biological or cultural preferences?

Two main elements have influenced social norms with regard to age differences in couples. The first is the gap in fertility ages between men and women and the second is the fact that human sexuality is characterized by long-term stable relationships, as detailed previously in this book. In societies in which sexual encounters are casual one-off events there is no reason for a man to prefer a younger woman to an older woman, assuming both of the women are fertile. In long-term monogamous relationships this is no longer the case. From the perspective of achieving maximal fertility, a man who is limited to one partner in a long-term relationship will prefer as young a woman as possible (assuming she is fertile) to ensure the greatest number of children over time.

Several years ago an interesting study was conducted in Finland in an effort to ascertain the optimal age difference between men and women in marriages for the purpose of maximizing the number of children reaching adulthood successfully.[9] The authors of the study based their findings on historical records in the Sami population between the seventeenth to nineteenth centuries. The Sami, native to northern Scandinavia, were chosen for this study, as was the range of years in the historical records, in order to identify the optimal age gap in an entirely natural environment unaffected by modern medicine. The researchers' conclusions grant Hugh Hefner and Woody Allen a bit of legitimacy in their choices of mates: the optimal age difference was greater than fifteen years. The couples included in the study exhibited a wage range of age gaps, from men who married women twenty years older to men who married women twenty-five years their juniors. The precise optimal age gap for having the greatest number of healthy children was found in couples in which husbands were 16.4 years older than their wives.

A follow-up study that focused on couples in modern Sweden, who obviously benefit from the latest in medical advances, concluded that the optimal age gap had narrowed to six years. But even in contemporary Western societies it is not extremely rare to find couples in which the age gap is as much as twenty years or more. It is no coincidence that large age gaps are particularly noticeable among celebrities and their wives, largely for social reasons. These couples usually form as a result of a "trade" in which the older man receives social credit for exhibiting vitality and youthful fervor while the younger woman receives in exchange social status, money, and fame. Television personality Larry King was once asked about the twenty-six year age difference between him and his wife Shawn.

"I know what you're thinking," replied King, "that when people look at Shawn and me, the first thing they notice is the age difference. But I'm here to say, if she dies, she dies."

Cliché 9: Men, more than women, seek physically attractive mates. Women, more than men, seek professionally successful mates.

The facts: Outward physical attractiveness in mates is important for both sexes to some extent, because in the past it was a marker indicating health and fertility. On the other hand, what is considered attractive in humans is very culturally specific and far from universal (except for some measures of facial symmetry, which have been shown to play a remarkably universal role in defining beauty). The attraction people feel to whatever is considered fashionably beautiful in a specific culture is mainly related to the social credit accruing from attaining a desirable mate in the mating market. That is also the reason that men tend to brag more than women about their romantic escapades.

When it comes to seeking professional success in mates, the picture is different. In prehistoric societies professional success was expressed in hunting skills. Good hunting skills improved the attractiveness of a mate in two ways. It increased the chances of feeding a large family of children. But perhaps even more importantly, since good hunting skills are to some extent inherited in successive generations, it also increased the chances of having successful offspring over several generations, a significant factor in improving the overall genetic success of the woman.

By this reasoning, a woman's professional success should also be important for men seeking mates. But men and women are not symmetric in their "reproductive strategies." As already noted, men have evolutionary grounds for stressing quantity (meaning absolute numbers of offspring) whereas women stress quality. This is apparently the evolutionary reason that professional success in mates is more important for women than it is for men.

Cliché 10: Women are more talkative than men.

The facts: Luann Brizendine published a book in 2006 titled *The Female Brain,* which provided the ultimate explanation for the ever-present tension so many of us experience in our marriages: women, claimed Brizendine, talk three times as much as men![10] Using data collected in her clinic, Brizendine concluded that women express on average 20,000 words per day compared to an equivalent male statistic of only 7,000 words per day. Brizendine compared the female brain to a highway for emotional processing. The male brain in this metaphor is more like a dirt road. These differences, she claimed, are due to the effects of testosterone, which cause men to think about sex so much that it blocks their ability to express emotions.

Though many men may think their wife or girlfriend is a shining example of this phenomenon, Brizendine is in fact wrong. About a year after she published her book, a thorough and wide-ranging research study was conducted on the subject of talkativeness in men and women by Matthew Mehl and several of his colleagues from the Department of Psychology at the University of Arizona.[11] Their conclusions were published in an article in the prestigious journal *Science.* According to the article, there is no difference in the number of words used by men and women. Members of both sexes express about 16,000 words per day on average. This finding was based on research using tape-recording devices attached to a large population sample. The three most talkative participants in the experiment were men, with the man topping the list clocking 47,000 words per day on average. The most important aspect here, however, is not the specifics of Brizendine's or Mehl's research claims, but the fact that so many people, male and female, are convinced (with slight exaggeration) that women are obsessive talkers, while men, in contrast, are imagined to be closer to monks who have taken vows of silence.

Why is there a large gap here between reality and the popular perception? One of the most talked about phenomena in the professional psychology literature on relationships is what is called demand/withdrawal, a situation in which one partner asks (or demands) to talk about problems in the relationship while the other partner seeks ways to avoid (or withdraw from) such conversations. Demand/withdrawal apparently plays an important role in the stigma of women as being overly talkative. A thorough study of the phenomenon conducted by UCLA researchers in 1990 showed that in most demand/withdrawal situations, it is women who are doing the demanding while men respond either with passivity or withdrawal.[12]

The UCLA study spurred efforts by several researchers to explain the different roles men and women play in demand/withdrawal situations. Some claimed that the demands expressed by women and the withdrawal exhibited by men were due to different ways in which men and women process emotions. But further research into the subject gradually showed that this is not the case. One such study showed that the active and passive roles in demand/withdrawal situations are determined not by the sex of the partners but by which partner initiates the conversation.[13] In general, the initiating partner is active while the other partner is the withdrawer, independently of the sex of the initiator. Another study, published in 2010, showed that demand/withdrawal occurs between homosexual partners (both male homosexuals and lesbians) to the same extent that it occurs among heterosexuals.[14] Finally, a 2006 paper showed that there is a significant cultural factor in the behavioral patterns exhibited in demand/withdrawal situations.[15] In Pakistani couples, for example, the roles we normally assign to men versus women are reversed. Taken together, these studies indicate that the role of women as the demanders in demand/withdrawal situations is not the result of different mechanisms for processing emotions. Apparently, women in relationships more often seek change, while men prefer to maintain the status quo.

This still leaves two questions relating to the phenomenon of demand/withdrawal. First of all, why do women seek change in relationships more than men? Secondly, why does the partner who is the receiver of the "demand" (whether man or woman) so often choose the passive "no comment" role, avoiding conflict instead of "firing back" or entering into a discussion of the issues raised by the demander? After all, we are usually very active in responding to demands we dislike when they come from people who are not our spouses.

We have, actually, already answered the first question when we discussed the evolutionary reason that women require emotional input and commitment from their male partners. Demand/withdrawal often occurs when one of the partners in a relationship (usually the woman) is complaining about a lack of commitment and involvement on the part of the other partner.

To answer the second question we will use some simple insights from game theory. Why does the partner who is the recipient of the demand withdraw? It has nothing to do with gender. In fact it has nothing to do with monogamous relationships at all. This phenomenon is typical of many intensive and long-term interactions. Demand/withdrawal is common, for example, between parents and their children (with the parent playing the role of the demander and the child withdrawing). The behavior associated with demand/withdrawal situations creates equilibrium between the partners to the interaction. In most cases demand/withdrawal is part of a long-running negotiation between the partners, where one of the partners is demanding a change that is costly to the other partner.

A strategy of passivity in response to demands does not necessarily imply lack of interest in the other partner's desires. Its goal is creating balance between accepting some of the demands without fully committing to all the changes asked for. In repeated interactions, this "no comment" strategy can be part of an equilibrium. Sometimes, silence is golden.

The last cliché relates to a commonly heard claim that is not directly connected to differences between the sexes but is relevant to the general subject of this chapter.

Cliché 11: Homosexuality endangers the survivability of the human race.

The facts: This claim is frequently raised by clergymen of nearly all the major religions who claim, "the creator favors the furtherance of the human race and homosexual sex can never be reproductive and therefore must be against god's will." Interestingly, some of the most secular individuals who believe in evolution would tend to argue the same, saying, "the forces of evolution favor the furtherance and reproduction and homosexual sex can never be reproductive."

This claim is true but reflects a shallow understanding of "kin selection," the evolutionary survival of a genetic line. The survival of the species is enhanced not only by those who produce new offspring but by those who ensure the survival of offspring with similar DNA (i.e., family members). Worker bees and ants, for instance, forgo reproduction and devote themselves to caring for the offspring of their queen, thus ensuring that her line continues to propagate. The phenomenon is not limited to social insects alone. The evolutionary advantage gained by avoiding direct sexual reproduction is that the "passive" individual, freed from the burden of caring for his or her offspring, can invest resources in caring for the offspring of others with a similar genetic inheritance, such as a brother or niece, thus increasing their chances of survival.

Empirical support for this explanation can be found in an article published in 2006 in the prestigious *Proceedings of the National Academy of Sciences of the United States of America.*[16] The article's authors reported that youths who have older brothers develop homosexual tendencies to a significantly greater extent than their peers who do not have older brothers. In addition, it was found that the same increased homosexual tendencies are not expressed when the older brothers are not biologically related (as a result of adoption or divorce and remarriage on the part of parents of the children). This implies that the source of the increased homosexual tendencies is biological, not social influences. (How a mother's second or third child might differ biologically from her first is not known; but, for instance, her eggs might change after her first has been fertilized.) In many cases the older brothers had children of their own to care for, and were helped in bearing this burden by their younger brothers.

Note that this explanation is independent of whether or not homosexuals in the modern world tend to care for their nieces and nephews more than heterosexuals; evolutionary pressures influenced these tendencies tens of thousands of years ago. It is therefore plausible that evolution selected for homosexuality in those cases in which having that trait conferred evolutionary advantages.

✳ ✳ ✳

DIFFERENCES BETWEEN THE SEXES DUE TO EVOLUTION WERE established at the dawn of human civilization, in a physical environment that was very different from the environment in which we live today. Many of those differences stopped giving individuals evolutionary advantages hundreds of years ago, if not earlier. Social mechanisms that adapted human characteristics to new environments undoubtedly changed many of the components of our personalities and our emotional reactions.

Why, then, have so many differences between the sexes continued to persist in our modern and advanced societies despite a century of feminism and policies directly intended to blur the distinctions? There may be more than one answer to that question, but I believe that there is one decisive answer: although some of the specific traits distinguishing men from women may not provide any evolutionary advantages in our day and age, the very fact that there are differences between the sexes is still a tremendous evolutionary advantage for the human race and for both sexes. Our sexuality is enhanced by the differences between men and women, encouraging sexual attraction and therefore reproduction. Men who blur their masculinity and women who blur their femininity reduce their chances of successfully competing for mates of the opposite sex.

Even in situations that have nothing to do with finding romantic partners, people consider stressing masculinity in men and femininity in women to be aesthetic and attractive. That is why instead of blurring gender differences in our outward appearances, we continue to stress them. Many of the most liberal and enlightened individuals in our societies still put on make-up (if they are women) or wear masculine-looking clothes (if they are men).

We often stress gender differences in behaviors for the same reason we stress them in outward appearances: consider how many couples you know in which the woman is more often in the driver's seat when both of them are in the car versus the number of couples in which it is the man who is more likely to be the driver. How about the number of men you know who do most of the cooking in their families versus the number of women?

Minor gender differences based on ancient evolutionary conditions can grow and become more prominent over time, instead of disappearing. That is why breast enlargement surgery became popular at a time when few women were breastfeeding their infants, and why men are still attracted to body-building gyms even though technological and economic advances have almost entirely eliminated the usefulness of physical strength in obtaining food and shelter for their children.

15

MAKE ME A MATCH MADE IN HEAVEN

Reproduction and the Mathematics of Romance

IT IS QUITE POSSIBLE THAT WE OWE THOSE WONDERFUL THINGS THAT we call love and sex to viruses. If viruses did not exist, all animals, including humans, would apparently reproduce asexually.

A large percentage of plant species and some animals do in fact reproduce asexually, meaning they do so without involving a second organism. But most complex species would be unable to survive viral infections if it were not for sexual reproduction. There is an unceasing war constantly taking place between animals and their virus enemies. The most effective weapon that animals have developed for winning this war is genetic variation.

Viruses attacking animals, including humans, are forever striving to adapt themselves to the genetic structures of their victims. Our genetic structures are analogous to locks, with the viruses looking for the corresponding keys to open those locks. Once they have those keys, they can attack every animal whose locks are similar enough to be opened by the keys. If there is sufficiently broad genetic variation in a given population, the viruses need to carry a very large bundle of keys in order to attack every individual. If, in contrast, a population is genetically identical, a virus with a single key can wipe it out entirely. Sexual reproduction enables two individuals with different genetic structures to mate and produce an offspring whose genetic structure is different from that of both parents. In essence, sexual reproduction is an insurance policy guaranteeing the genetic future of the parents.

This is also the source of the evolutionary taboo against sexual relations involving family relatives. It might seem that if the goal of evolution is to create a hereditary chain of animals sharing as much genetic similarity as possible, natural selection would favor reproduction within the family unit, with brothers and sisters as optimal mates for producing children. In fact, incestuous reproduction is a sweeping evolutionary liability.

There are genetically transmitted diseases whose incidence is greatly amplified in children produced by incestuous relations. In addition to the social taboo against such relations, we have also developed an efficient psychological mechanism that prevents us from being attracted to close family relatives. All of this protects the genetic variation in our species, even at the price of reducing the genetic nearness between ourselves and our offspring. A tight genetic similarity between us and our offspring means less variation in the human population, making our species more vulnerable to extinction at the hands of a viral plague.

This is, of course, another example of emotions being a hugely valuable mechanism for preventing a bad outcome: the logic required to understand the evolutionary risks of incest is far more abstract than the immediacy of our revulsion to it. Nearly all of us recoil at the thought of conducting sexual relations with family relatives such as siblings or cousins, but many studies show that most of us are in fact sexually attracted to those who resemble us both in appearance and personality. Psychologists who have conducted studies of the phenomenon have found that siblings and cousins who are not aware of the familial ties between them (as can happen in cases of adoption, separation of parents, or very large families) report significantly greater sexual attraction between them relative to most couples. It is reasonable to suppose that this attraction stems from the fact that if the viral threat were not present, there would be considerable evolutionary advantage to marrying relatives.

It is interesting to imagine a science fictional account of a world without viruses in which humans reproduce asexually. If there were no need to fight viruses by confusing them with genetic variation, there would definitely be an evolutionary advantage to asexual reproduction. Sexual reproduction is genetically inefficient: it is complicated, leaves too much to fate, and above all it produces offspring who are not genetically identical to their parents. Asexual reproduction, in contrast, would have enabled us all to clone ourselves and form perfect genetic copies of ourselves. It is reasonable to assume, based purely on evolutionary considerations, that if we lived in an asexually reproducing world, asexual reproduction would be as pleasurable to us as we find sexual relations to be. If reproduction were not pleasurable we would not do it, and then we would cease to exist as a species.

Evolutionarily, we could probably get along fine with asexual reproduction, but what about human society in such a world? What would play the roles of courtship, love, romance, and flirting? What would art and music be like without the subject of romantic love? Narcissism and egocentrism would doubtless be primary personality characteristics in an asexually reproducing world. We would all be focused on ourselves, rarely interacting with others. In sum total, our lives would probably be emotionally poorer and far more boring.

But if two are better than one when it comes to sex and reproduction, why aren't three better than two? This question was posed and studied by Motty Perry and two colleagues in a very interesting research article.[1] Given that sexual coupling is a way to ensure genetic variation in a population as protection against viruses, why didn't natural selection go on to create threesome sexual reproduction? After all, by combining the genetic material of three individuals, even more variation would be created.

I should point out that by threesome sexual reproduction I do not mean ménage a trois scenarios as one might find in certain French genre films from the 1970s—two men and a woman sharing a bed, or two women and a man. A world with threesome reproduction means a world with three different sexes: male, female, and a third sex for which we obviously do not even have a name in our languages. In such a world, each successful sexual encounter would require the involvement of one member from each of the three sexes, each contributing genetic material for the creation of offspring. There is no species on earth that reproduces in such a way. There is good reason for this: the advantages that threesome sexual reproduction has relative to the sexual coupling reproductive method we are so familiar with are overwhelmed by the disadvantages it brings.

From a technical perspective, there is no difficulty at all in imagining threesome reproduction occurring. There are cases in which DNA tests for the purposes of establishing paternity have mysteriously failed to show a genetic connection to either the father or the mother. In some of those cases, further investigation revealed that the child in question actually had three parents, as the result of an ovum that was fertilized by the spermatozoa of two different men, with the child then bearing the genetic heritage of both men and the mother. The child's mother, it turned out, had indeed engaged in sexual relations with two different men in a short space of time and her ovulating egg had been penetrated by the sperm of both men.

Perry and his coauthors showed that adding more distinct sexes in the act of procreation does indeed lead to greater genetic diversity in a population, but the increase in diversity gained by moving from two sexes to three sexes is marginal. On the other hand, reproduction requiring three (or more) distinct sexes reduces fertility significantly, because it requires three individuals desirous of reproduction to find each other and meet, which is much more complicated than the analogous chance meeting of two individuals. The conclusion is that reproduction in pairs is the optimal form of reproduction for organisms seeking to avoid being wiped out by viruses. It is comforting to know that the human approach to sexuality through a romantic attachment to one other individual, as opposed to groups of three or more, is not arbitrary but the result of mathematically well-founded evolutionary considerations.

Previous chapters of this book pointed out that human sexuality differs from the sexuality of most other animals because it is based on a mix of emotions and commitment. But the emotions that so move us in romance and sexuality are not arbitrary either. In contrast to common perceptions, we do not suddenly fall in love or get swept away by romantic emotions; love develops at the right time and with the right person. In fact, love is in most cases the result of decisions we make.

When I completed my university studies and went to the United States as a young researcher, I was astonished to discover that some of my colleagues, researchers from India, had married their spouses by way of arranged marriages. These colleagues were young, modern, and liberal in their views and were extremely well educated and intelligent. They had lived in the United States for many years, but when it came to marriage, they accepted the traditions of their culture and entered into matrimony in matches arranged by their parents.

When I spoke at length with my Indian friends on the subject of love and relations between the sexes, they described their experiences of growing to love their spouses as the result of rational and deliberate decisions that they took. When they had met their spouses-to-be for the first time, a wedding date had already been set, as had their future living arrangements and the amount of the dowry that the father of the bride was to pay the father of the groom. The question of whether or not the bride and groom were appropriate for each other as a couple remained almost entirely in the hands of the parents.

In the Indian arranged marriage tradition, during negotiations on the amount of the dowry both the virtues and the negative aspects of the bride and groom are openly discussed. If the gap in the "qualities" of the members of the proposed couple is regarded as too large by the parents, negotiations are ended and the respective parents will seek new matches for their children. Small gaps are "bridged" by appropriate tweaks to the amount of the dowry, reflecting the relative qualities of the children whose marriage is being arranged.

A colleague of mine named Ragavan went to India to get married in the middle of his doctoral studies at the University of Oxford. He got to meet his future wife for half an hour before the marriage agreement was concluded between their parents. Within two days the newlyweds were on their way to Oxford. Ragavan and others have often told me that the process by which their marriages were arranged has not reduced in the least the love that they feel for their spouses. In fact, they claim that quite the opposite effect is achieved. They were able to concentrate on building a loving relationship after all the other details regarding their marriage had been determined, arranged, and completed. Some of my Indian friends have even told me that they have difficulty understanding me and my wife. How could both of us start dealing with such an emotional matter as a loving relationship, they ask, when there was still so much uncertainty surrounding it?

For all its positive aspects, the system of arranged marriages as practiced in India (and many other countries), and especially the custom of paying dowries for marriages, can also be the source of several social ills, most prominently inequality between the sexes. The dowries that the father of the bride is required to pay the father of the groom can be beyond what the bride's family can bear. There are Web sites in India containing dowry price lists. The price is mainly determined by variables such as the groom's occupation, his caste, and the caste of the bride. Large gaps between the standing of the bride and groom can raise the price of a dowry to upwards of $130,000. It is, therefore, not surprising that the birth of a daughter is regarded as a burden in many Indian families, while the birth of a son is treated as a treasure. In recent decades, with the development of technologies that can identify the sex of a fetus in the early stages of gestation, a trend toward aborting female fetuses has gained momentum in India (and in China as well). Prior to the trend of female fetus abortion, biology gave the general human population an identical proportion of women and men. The abortion of female fetuses has changed that, with men now outnumbering women in the world population by 2 percent. In India the gap between men and women in the population stands at 4 percent more men, while in China that gap is now 6 percent. In some provinces of India the gap is even wider. Interestingly it is in wealthy areas that some of the widest gaps have emerged, because wealthy women carrying female fetuses are more able to bear the cost of having an abortion than poor women.

These imbalances have inevitably set in motion correcting market forces. A shortage of women has brought about significant reductions in dowry prices. In some places, in a reversal of tradition the parents of brides are now demanding that the groom's parents pay them a dowry for the right to marry their daughter. In areas with especially large gaps between the numbers of men and women, the shortage of nubile women is so great that another perturbing economic phenomenon has emerged: two brothers marrying the same woman to enable their family to pay for the large dowry demanded by the bride's parents.

The dating market in Western societies is freer and more spontaneous, but a careful look at the rational and economic considerations operating in it shows that it is not that different from the traditional arranged marriage system customary in India. The phrase "love is blind" sounds poetic, but reality is usually much more prosaic. In most cases we fall in love with individuals with whom we expect we can form mutual bonds, while avoiding developing feelings of love for those whom we believe are "unattainable." Romantic attachments often form between couples belonging to the same ethnic group and sharing the same social and economic standing.

My colleague and friend Eva Illouz has conducted a thorough study of the way women and men choose romantic mates in modern Western societies.[2] Illouz's study shows the extent to which liberalism with regard to relationships, along with technological advances that now enable romantic meetings to be arranged at the press of a button, have brought capitalist consumer culture into our love lives. When it comes to modern love, it turns out, we refuse to compromise on anything less than the best deal we can obtain, just as we do when we go shopping. To obtain that idealized goal we are willing to endure hundreds of Web-enabled dates that end in frustration and disillusionment. As a result we often avoid investing the commitment that is required for building stable relationships.

Gary Becker, winner of the 1992 Nobel Prize in economics, has also claimed that the decisions we make with regard to relationships and love closely resemble our decisions in market situations. In two articles he published in 1973 and 1974, entitled "A Theory of Marriage," Becker presented a mathematical model of the marriage market.[3,4] He was not the first to do so. A decade earlier David Gale and Lloyd Shapley, mathematicians specializing in game theory (Shapley was awarded the 2012 Nobel Prize in economics), fashioned a similar model for the marriage market.[5]

Both models describe a two-sided market with women forming one side of the market and men the other. Each man lists the women in order of preference: the women to whom he is more attracted are at the top of the list, while those to whom he is less attracted are toward the bottom of the list. Each woman has a similar ranking of the men, from most attractive in her eyes to least attractive. Each market participant, male or female, retains the right to remain single if the only members of the opposite sex that are available to him or her appear so low in his or her subjective attractiveness ranking that being alone seems preferable to marriage.

The central concept underlying both models is the beautiful idea of a stable set of matches. A set of matches is a monogamous assignment of the men in the market to the women in the market. Each man is matched with no more than one woman (although some men may remain single, i.e., not matched with any woman) and each woman is matched with no more than one man (again, some women may remain single). A set of matches is stable if it is not possible for any couples to divorce, or for the people involved to be better matched than they already are. Man A may desire Woman B, but if she also desires him and they are each married to someone else, it is not a stable set of matches. Similarly, in a stable set of matches each individual who was matched to a member of the opposite sex prefers being with his or her spouse to remaining single.

From these definitions alone, it is not immediately clear that such an ideal stable set of matches can always be constructed given a marriage market of men and women. Using a short and very elegant proof, however, Gale and Shapley proved an optimistic mathematical theorem: a stable set of matches always exists—no matter what the preferences of the men and women in the marriage market are! Gale and Shapley even showed how a simple and easily implementable procedure, that can be run on a computer, can find a stable set of matches, given an input of the preferences of each man and each woman.

Gale and Shapley's model, more than Becker's model, has turned out to have very broad applications. In fact it is one of the most influential and practically applied economic models of all times. It has been used, for instance, to find placement for medical interns, a market that got dramatically more efficient as a result. And, after the esteemed Stanford economist Alvin Roth further developed the theory, it has helped school boards in the United States and the United Kingdom improve the placement of children in their preferred schools. In recent years Roth has also been the driving force behind the introduction of the Gale-Shapley algorithm for a novel application that is literally life-saving: kidney transplants.

Successfully implanting a kidney requires a high level of genetic compatibility between donor and recipient. Not unlike marriages, many potential implants never happen because the two are not (genetically) compatible. Alvin Roth, along with several colleagues, had the insight to recognize that many lives could be saved by using matching algorithms to bring together compatible donors and recipients. The idea goes as follows: suppose that Ron wants to donate a kidney to his ailing sister Ruth, but unfortunately their compatibility is insufficient for a successful transplant, while at the same time Maya wants to donate a kidney to her husband Gary, with that transplant also vetoed by doctors due to incompatibility. If Ron's kidney, however, can be successfully transplanted to Gary, while Maya's kidney is compatible with Ruth's body, an organ "swap" between the pairs can be conducted, saving two lives that might otherwise be lost.

Roth correctly understood that the potential "market" between such organ donors and recipients is similar to the marriage market and the interns market mentioned above: it is a two-sided market with donors on one side and patients in need of kidney transplants on the other side. An algorithm can therefore be applied in the transplant market to create long chains of kidney donors and recipients, saving thousands of lives annually across the country.

Yet the Becker model has much value of its own: it illustrates the surprising savvy and self-interest embedded in our decidedly nonrational dating market. Here's how it works. In Becker's model, people rank the attractiveness of members of the opposite sex using a set of characteristics such as appearance, education, social standing, wealth, and so on. Each individual gives different relative weights to these characteristics in forming his or her preferences. Each potential match between a man and a woman creates "joint utility," a jargony term meaning the benefit each partner gains from the match, based on the characteristics of each and the weights that the other assigns to those characteristics. A more successful couple has higher "joint utility," although it is not necessarily the case that they share it equally— more about that in a moment. In contrast to the Gale-Shapley model, which permits individuals only to agree or refuse any particular match suggested to them, in Becker's model the members of every couple that is formed also need to agree how to divide between themselves the joint utility that their match creates.

Consider, for example, a woman with many attractive characteristics who is rated highly by many men. It is possible for her to marry a man who is not considered attractive by other women. But in that case the division of the joint utility generated by the marriage will be skewed in the woman's favor. This may be expressed, for example, by the man having to do more of the household work or forgoing the purchase of a sports car that he particularly desires. This assumption is called "transfer of utility" in Becker's model and it is the essential difference between Becker's model and that of Gale and Shapley. The extent to which the assumption of transfer of utility is reasonable is a subject of ongoing dispute between economists, and we shall return to it.

Here is an example of how a stable system of matches and agreements on utility transfers may be accomplished under Becker's model. To simplify matters, the marriage market in the example is small, containing only two women, Rachel and Miriam, and two men, Sam and David. Each of the four possible matches that can be formed in this marriage market creates joint utility for the couples involved. These joint utilities are illustrated in the following table:

	David	Sam
Rachel	8	4
Miriam	9	7

As the table shows, if, for example, David and Rachel are matched together, their joint utility as a couple is 8, representing the benefits that the couple gets from being a couple. This includes both the material and the emotional benefits that both members of the couple get from the fact that they have married.

Note, though, that in our example, although David and Miriam are the most successful potential couple (creating a joint utility of 9, higher than all the other utilities in the table), they cannot be matched in a stable system of matches. To see why, consider the situation in which David-Miriam and Rachel-Sam are the matched couples. Suppose that Sam, David, Rachel, and Miriam individually receive the utilities S (for Sam), D (for David), R (for Rachel) and M (for Miriam), respectively, under the utility division agreements that the couples sign. This means that $S + R = 4$ and $D + M = 9$. Put simply, Sam and Rachel are a bad couple: based on our chart, he would be better off with Miriam and she better off with David. Whatever they are getting in their current match, they can get more by being together and dividing the pie of 8 units. Thus, even though David and Miriam are a happy couple, the unhappiness of Sam and Rachel makes this set of matches unstable. They have an incentive to divorce. Another way to look at this is to note that the sum of the couples' "utility" is greater in a stable set of matches. Sam and Rachel's joint value is 4, and David and Miriam's is 9, for a total of 13; on the other hand, David and Rachel's utility is 8, and Sam and Miriam's is 7, for a total of 15. The greater number, in this case, reflects stability.

The important insight here is that the stability of a couple depends on more than the direct relationship between the two members of the couple—it also depends on what is possible outside of the partnership, meaning the possibility that each member of the couple can do better by finding an alternative spouse. For the same reason, it is entirely possible that the most successful relationship (meaning the relationship that creates the greatest amount of joint utility) will never be part of a stable system of matches. For a system of matches to be stable it must maximize the sum total of the utilities of *all* the individuals in the marriage market.

In our example the system of matches composed of David-Rachel and Miriam-Sam is stable. It creates a sum total of 15 units of utility, the greatest amount of total utility possible in this marriage market. The next interesting question is then how will these utilities be divided between the man and the woman in each couple? The answer again depends on the entire market, and it will not necessarily be divided equally between the man and the woman.

Suppose that the utilities were divided equally in our example, so that David and Rachel agree to divide their joint utility of 8 in a 4:4 split while Sam and Miriam agree to a 3.5:3.5 split. This is an unstable arrangement, because Miriam and David could divorce their spouses and form a new couple enabling them to divide more units of utility between themselves (9 units instead of 7.5). There is a division of utilities in this case that does lead to a stable system of matches, as follows: David and Rachel divide their 8 units of utility equally between them at 4 apiece, while Sam and Miriam divide their utility of 7 by giving Sam only 2 units of utility and Miriam 5—Miriam thus receiving three more units than her spouse Sam.

Under such an arrangement, why wouldn't Sam rebel against the inequality imposed on him and demand, for example, that Miriam handle the laundry in their household, given that he is already doing all the cooking and driving the children to soccer practice? The (cynical) answer given by Becker's model to this question is that if Sam does receive a larger share of the joint utility at Miriam's expense, then Miriam will have an incentive to divorce Sam and marry David instead, in a new arrangement that would make both Miriam and David better off.

If you find the materialism, unmitigated self-interest, and selfishness that pervades Becker's model somewhat unpalatable when the subject matter is relationships and love, I sympathize with you. But let's be precise in our criticism of the model. Becker's model isn't necessarily talking about pure materialism, since as mentioned above the numerical values in the model also represent emotional utilities. But the model is indeed based on self-interest and selfishness, which is one of its weaknesses. The model implies, for example, that if one member of a couple is injured to the extent that the joint utility of the couple is significantly reduced (such as might happen if there is a serious illness or a similar disaster) then his or her spouse should immediately start seeking a new relationship. This is not an accurate portrayal of a loving relationship—not from a moral perspective or an empirical one.

Gary Becker proposed his model of the marriage market in the 1970s, when he was a member of the economics faculty of the University of Chicago. The "Chicago School," as it was known, was marked by an approach to economics characterized by material self-interest on the part of economic agents and an unequivocal belief in the forces of the free market. Becker was very much a supporter of this approach. Given this, it is not surprising that Becker also controversially proposed that human organs be bought and sold on the free market for the sake of reducing chronic shortages of organs for transplants.

Despite the criticisms that can be leveled against Becker's model, it is still a very important model because it provides us with several insights into how the marriage market actually works. Some of these insights are well supported empirically. The model, for example, correctly predicts that increasing the participation of women in the labor market can improve the standing women have within their relationships with their partners but can also increase the frequency of divorce. This follows from the fact that a woman's utility from staying single increases once she has the means of supporting herself.

Consider what would happen in the above example if men and women can remain single. If each person's utility from staying single is 1, nothing would change in our example (since being part of a couple delivers each person more than that). Suppose now that the opportunity to work and support herself were to increase Rachel's utility from being single to 4.5, up from 1. Then the system of matches previously detailed would no longer be stable. Rachel would demand at least 4.5 utility units from David in exchange for remaining in a relationship with him, leaving David with only 3.5 (out of 8). But David need not agree to Rachel's demand. He may instead offer Miriam 5.2 units of utility if she agrees to marry him (which is more than she is getting from Sam), which would make both of them better off. This analysis provides some support for the claim sometimes raised by activists in women's organizations that many men are opposed to their wives working outside of the house not because of concern that this will reduce the quality of care their children are getting, or because of fears that household work won't get done, but because they are worried that economic independence for women will increase women's bargaining position within relationships.

Another weakness of Becker's model is the assumption that utility can be transferred. By this assumption, just about any negative characteristic that one member of a couple sees in the other can be corrected by dividing the joint resources in the marriage in an appropriate way. This sort of brutal smashing of the entire basis of romantic attachment in relationships is not only outrageous, it is also not an accurate portrayal of reality. Rather than speaking in anyone else's name, let me relate a personal story. When I was a freshman student, I had a brief romantic relationship with a woman who had just about everything I could ask for. She was good looking and intelligent and had an amazing sense of humor and great sensitivity. But what I could not find in her, despite trying very hard to find it, was that amorphous and ill-defined thing that distinguishes a very good friendship from a sweeping sense of falling in love. I cannot imagine that this exquisite young woman could have given me anything at all to compensate for that missing ingredient.

16

FROM CAVEMEN FLUTES TO BACH FUGUES

Why Did Evolution Create Art?

MY FATHER, WHO PASSED AWAY SEVERAL YEARS AGO, SOMETIMES VISITS me in my dreams in the middle of the night. One day, musing over the memory of such a dream from the previous night, I grabbed a pencil and paper and composed a poem to my father that I later set to music. As I was writing, the words flowed onto the page with an ease I had never previously experienced. My writing is usually hesitant and I tend to make endless corrections and changes to the words. That wasn't the case here. The tune also came to me very easily, but when the composition was complete and I cradled my guitar in my hands to play it, my eyes welled with tears. I was so overcome with emotion that I simply could not sing the words.

That's narcissistic, I thought to myself for a moment, getting choked with emotion over your own composition. But I immediately understood that it wasn't the lyrical quality of the song or the love of my own creation that were bringing on these emotions. Neither were longings for my father ultimately causing them. I was moved mostly by the accurate way the words of the song described my father as I imagined him.

Although the words flowed easily when I wrote them, they still required much intellectual effort to compose. I recall the experience as being similar to what I go through in my daily research, which mainly involves proving mathematical theorems. The exact rhyming of the words and the unusual tempo were part and parcel of the strong rush of emotions. Where did the song come from and develop? Was it in my head or in the depths of my heart?

We tend to think of our emotions and analytical thoughts as arising in two separate internal systems. Under the best conditions, we hope that those two systems will not interfere with each other. In less favorable cases, we fear that they will become locked in an implacable struggle. The truth is apparently very different from this description. There is a thin border between our emotional and analytical/cognitive internal systems. An intense dialogue between the systems takes place mainly in the prefrontal cortex, which is located at the front of the brain, protected by the forehead.

One of the most successful contemporary treatments of clinical depression involves placing magnets near the prefrontal cortex to undo the characteristic vicious cycle of negative thoughts creating ever more depression and anxiety. But there are also positively reinforcing cycles of thought taking place between the emotional and cognitive systems. In fact, every experience with art, whether it involves the creation of new art or its appreciation by an observer, involves such a dialogue. Artistic experiences are clearly related to emotional responses, but those emotional responses are created within a cognitive process in which we seek to arrive at an insight or to identify an aesthetic structure in the work of art.

Nearly every artistic experience is a combination of cognitive analysis and emotional reactions. Without the emotional part we would remain unmoved and indifferent to the artistic creation. It would seem unimportant to us. But without some level of cognitive analysis we would be unable to identify the aesthetic qualities of the creation, and then it would be incapable of arousing an emotional reaction. Artistic creations that aim to elicit visceral emotional reactions (such as extremely violent motifs or heart-rending descriptions of suffering) are usually regarded as shallow and incapable of arousing true artistic experiences.

Johann Sebastian Bach, whom I regard as one of the most emotionally moving composers of all time, was also a very mathematical composer. Bach's fugues are almost entirely bereft of a main melody, but they create an intricate web of sounds of different tones combining in incredible sophistication into a giant puzzle. The *Fugue in C-Minor* is constructed entirely out of four voices, each of which moves along an unconventional 1:16 time signature, together with an elongated pedal tone. The fugue is based on the so-called "BACH motif." It uses the notes that together compose the name of B-A-C-H. (Many music scales go from A to G, but in some cases H is considered a sharp version of B.)

I recommend that you view an online video of people showing off how rapidly they can solve Rubik's Cube challenges. If you play a Bach fugue as a soundtrack to these videos, you will soon see that they fit perfectly, as if the fugue is guiding the people in the videos toward the desired solution.

But where do artistic experiences come from? What ends, if any, are served by this synthesis of emotion and logic?

Several years ago a twenty-centimeter flute carved out of an eagle's bones was discovered in a cave in southern Germany. This flute is considered to be the oldest known musical instrument created by humans. Scientific analysis revealed it to be 35,000 years old. The oldest known cave paintings are dated to about the same time period. Artistic creativity apparently predated most of humanity's cognitive development. It goes back to the beginnings of human evolution. The pleasure we take from art and our need to experience it may be relics of an ancient human survival need for connecting with others.

Neurobiologists have been trying to understand why the brain reacts so emotionally to music, to the point that it can raise goose bumps on the skin. In one study a few years ago subjects were asked to select their favorite musical passages (from purely symphonic works with no accompanying words). Those musical passages were then played for them while fMRI scans of their brains were conducted. The brain region revealing the greatest activity in those scans was the striatum, a subcortical brain region that is responsible for the secretion of dopamine—a hormone that is involved in giving us pleasurable feelings in a wide range of contexts, including sexual activity and the temporary highs brought about by some addictive drugs.

The way that music affects our emotional moods is fascinating. Music can have dramatic effects on our minds. We tend to enjoy hearing musical passages with familiar structures, but the familiarity cannot be too great, because then we get bored. Brief moments in which surprising sounds emerge in contrast with more familiar and expected notes bring the greatest pleasure. In other words, we apparently require the anchor of the familiar to enjoy the unfamiliar.

There is a common denominator tying together the pleasures we get from music on one hand and well-done humor and successful jokes on the other hand. In both cases pleasure comes from the contrast between expectation and surprise. The same holds for gripping passages in books and films.

Taking pleasure from surprises begins very early in infant development. Infants as young as a few months old are easily prompted to hysterical laughter by the sight of a familiar person doing something unexpected, while we, watching from the side, are honestly no less moved by watching this sight. A good example of this can be seen in the following YouTube video, which has attracted more than 45 million views:

The baby in this video gets incredible pleasure from watching a family member rip up pieces of paper. Why? Because the paper ripping is a surprise to her. Why do we take so much pleasure from surprises? Do our emotional reactions to surprises give us any survival advantages? The answer is that we learn to recognize our physical and social environments mainly through surprising experiences. Each surprising experience implants important knowledge in our brains that can be drawn upon in the future for better decision making.

Familiar experiences rapidly disappear into black holes of forgetfulness. That is a good thing, because they provide us with knowledge that we already possess. Surprising experiences can give us new and vital information. The emotional pleasures we get from surprising experiences incentivize us to seek them and to be alert to their existence, thus improving our learning and our chances for survival. There are other mechanisms that boost learning, such as curiosity. But these mechanisms are more cognitive and therefore are slower than the emotional mechanisms prompted by music and humor.

But we need the structure of the familiar to enable us to learn from surprising experiences. A world composed entirely of surprises would be one in which we could never learn anything. It would be unfamiliar and strange. We would not be able to see ourselves as part of such a world, nor would we know what to expect in the future based on past events. This is why music based on tempos and scales that are unfamiliar to our ears sounds cacophonous. A film showing only a procession of surprising scenes will look tiresome and strange. A stranger trying to surprise an infant by covering and then uncovering his or her face can sometimes induce anxiety instead of pleasure in the child, leading to frightened tears in place of rolling laughter.

PART IV

On Optimism, Pessimism, and Group Behavior

17

WHY ARE WE SO NEGATIVE?

The Arithmetic of Emotions

IMAGINE DISCOVERING ONE FINE DAY THAT YOU ARE IN POSSESSION OF a winning lottery ticket that gives you an instant prize of $100,000. You literally jump for joy.

Now imagine that a week later you buy another lottery ticket, which amazingly turns out to be a winning ticket, giving you yet another $100,000. Another week later, the same thing happens.

Try to rank the levels of joy you would feel each successive time you won the lottery. When would you be happiest? If your answer is that you expect your joy would be strongest the first time, with less additional joy the second and third times, you are no different from most people. Your intuitive expectations in this scenario correspond to one of the most basic assumptions in economic theory, known as "diminishing marginal utility," which posits that the more wealth you have, the less each additional dollar (or $100,000) adds to your welfare. The marginal utility gives us the extent by which our welfare goes up (or down) as our wealth increases (or decreases).

Diminishing marginal utility fits well with our common sense observations. Giving a struggling student barely able to make ends meet $100,000 will likely have a dramatic effect and elicit a great deal of joy. Giving that same $100,000 to someone as wealthy as Bill Gates probably would not have the same outcome—it would be financially meaningless to him and not change his mood in any way.

What happens with negative events is less obvious. Most of the evidence collected by behavioral economists suggests that the suffering endured from, say, losing $2,000 is less than twice the suffering endured from losing only $1,000. It is hard to prove whether the same applies when more serious negative events are involved, i.e., in case of a death of a loved one or an illness, but most economists tend to believe that this is indeed the case. In pseudo-arithmetic terms we can paraphrase this as: one plus one adds up to less than two, but minus one plus minus one adds up to more than minus two.

How does the arithmetic of emotions work? There is very little scientific research on the subject to date. Theoretical economics gives a partial answer to the question with the concept of utility functions. A utility function associates each situation (where "situation" may mean a basket of goods, winning a particular prize in the lottery, or even contracting a disease or suffering personal injury) with a numerical value. This numerical value is meant to represent the subjective emotional reaction of an individual to each such situation.

In 1944, the mathematician John von Neumann and the economist Oskar Morgenstern published one of the most intellectually important books of the twentieth century, *The Theory of Games and Economic Behavior*.[1] In their book, von Neumann and Morgenstern studied utility functions in great detail. One of the more elegant results that they achieved was showing that a person who reacts with decreasing marginal joy to good news will be risk averse.

If offered a choice between a risky lottery or a risk-free sum of money equal to the average payoff of that same lottery, a risk-averse person will always prefer the risk-free option. For example, suppose you offer someone a choice between $1,000 with no risk or a lottery ticket with a 50 percent chance of winning $2,000 and a 50 percent chance of getting nothing. A risk-averse individual will prefer the $1,000 "sure-thing" option, despite the fact that the lottery offers a chance at winning $2,000.

Most people are risk averse. That is why insurance companies earn fortunes in profits. Most of us will choose risky investment instruments, like stocks, only if the expected average payments they offer are higher than those offered by more solid investments. It is true that many of us also like to buy the occasional lottery ticket or sometimes try our luck at the tables at Las Vegas casinos, but this sort of seemingly "risk-seeking" behavior usually involves relatively small amounts of money and can be categorized more as entertainment than true risk taking (unless it gets out of hand and turns into a gambling addiction, which is another matter that we will discuss later in the book).

Our attitudes toward risk are not immediately self-evident from an evolutionary perspective. There are animals that exhibit different risk attitudes from humans. A coauthor of mine, John Kagel, who is a leading researcher in behavioral economics, started off his research career by studying how pigeons relate to risk. Kagel, along with a group of other researchers, conducted experiments in which pigeons were exposed to several different pigeon holes holding differing amounts of food.[2] Some of the pigeon holes always had the same amount of food, while others had food amounts that varied over time. The distribution of food in each pigeon hole over time was carefully controlled so that on average they all had the same amount of food.

For example, if one pigeon hole always contained 20 grams of sunflower seeds, another pigeon hole might have 40 grams of sunflower seeds 50 percent of the time and be empty 50 percent of the time.

In contrast to the risk-averse behavior that humans typically exhibit, the pigeons preferred pigeon holes with random amounts of food to those that always contained the same amounts. Kagel suggested that the difference in risk attitude between humans and pigeons might be based on the different environments in which the two species live. Pigeons need a minimal amount of food to survive. A food source that provides less than the minimal amount that a pigeon needs is of no use for surviving. The regular sources of food that pigeons usually encounter in the wild might not provide the minimally necessary amounts, leading pigeons to prefer taking risks in the hopes of obtaining food amounts greater than the minimal threshold they require.

The consumption environment of humans is arguably very different from that of pigeons. Think of how your welfare is affected by different quantities of some commodity that you own. An extra unit from the commodity will raise your welfare substantially if you own little of it, but the rise in welfare will be much less pronounced if you already own a lot of it. In essence, this is what makes us risk averse. Why? Imagine you own five apples and I offer you to trade these apples for a coin toss: if it turns up heads, I give you five more apples; tails, I give you zero apples. If you make this trade, you either lose five apples or gain five apples. If apples are all you have to eat, losing five apples reduces your welfare (because you might starve) more than gaining five apples raises it (because you can try to stretch the five apples longer). Hence, you'd be better off not accepting my offer but rather remaining with your five certain apples. In other words, risk aversion is a rational trait for us humans, and this is why we almost consistently display this trait.

So far we have considered the arithmetic of emotions as it relates to similar events. But what happens when we take into account completely different events? How do winning the lottery and a great night out add up emotionally? How does being informed that you have been awarded an important promotion at work balance against news that a close friend has suddenly died?

Very little research has been conducted to answer these questions, and most of what we know on the subject is indirect. We know that our emotional reactions to events—whether positive or negative—are strongly influenced by the extent to which we focus cognitively on one event or another. If, for example, we experience two positive but different events in quick succession, limits to the amount of attention we can give to both events usually lead us to focus more on one event than another, with more emphasis generally given to the event that we regard as more important. This tends to reduce the cumulative effect of the other event on our emotional state. The emotional reaction will then be close to the maximal reaction we would have to only one or the other event. The emotional arithmetic here is therefore different from the simple addition of two joyous reactions.

The same thing happens with regard to two negative events. Our cognitive attention will be given to the worse of the two, causing the emotional effect of the other negative event to be marginal.

The more interesting scenario is one in which one event is positive and the other event is negative. In this case, again, the relative importance we attach to each event greatly determines which event will have greater effect on our emotional state, but, unfortunately, negative events almost always dominate positive ones. In other words, to cause us to focus on a positive event in juxtaposition with a negative event, the positive event needs to be regarded as much more important than the negative event. If it is only slightly more subjectively important, we will focus more on the negative event and the net emotional effect of both events will be negative.

Clinical depression is frequently accompanied by an obsessive focus on negative thoughts that almost entirely crowd out positive thoughts. Most of us don't experience extreme focus on negative thought to such an extent that we suffer from clinical depression, but unfortunately when it comes to the addition and subtraction of joy and sadness, even the healthiest among us tend to give more weight to the sadness.

ON ARROGANCE AND HUMILITY

The Norwegian Professor's Syndrome

WHILE THE INTREPID BIOLOGICAL RESEARCHER AMOTZ ZAHAVI WAS HARD at work developing his theory of the handicap principle, another researcher, the economist Michael Spence, was working on a similar idea that eventually won him the Nobel Prize in economics. That idea is called "market signaling."

Spence originally wanted to understand why people make such an effort to obtain college degrees before entering the employment market, even in the many cases in which the specific subjects that they studied in college gave them no directly discernible skills that were needed for the jobs they eventually took.[1] Spence's explanation is based on the fact that different people have different intellectual talents, and it is those talents that are the most important predictors of job success—much more than the content of their education.

Spence assumed that people with strong intellectual talents require less effort to complete a college degree than their colleagues who lag behind on the intellectual scale. When an individual applies for a job,

his degrees document how much schooling he has had, but not necessarily his true intellectual capacities. This leads to a market situation
in which the highly talented accumulate a large number of years of
schooling in order to broadcast their intellectual capacities. In other
words they "signal" their intellectual advantages to the market by
way of the number of years they have chosen to remain in the formal
educational system.

This signal is well understood by employers as proof of their intellectual talents, and degrees are thus translated into higher salaries.
The effort that the less intellectually gifted need to make to attain
higher education degrees is, according to this theory, so high that
it overshadows the higher salaries they could earn if they artificially
increase their years of schooling. In this way the market manages to
distinguish the intellectually talented from the others without making
everyone take IQ tests. The system of higher education indirectly
does the task for them.

There is a clear relationship between Spence's market signaling and
the handicap principle. The more intelligent individuals in our society
accept the "burden" of more schooling (I have students who readily
identify with the idea that taking classes constitutes a burden . . .) because they know that their weaker competitors cannot successfully cope
with this burden.

Spence's market-signaling model has been greatly expanded over
the years, and it is now used to explain many different economic phenomena. For example, why do manufacturers offer warranties for
their products? Because only the producers of quality products can
take on the monetary risks inherent in issuing warranties. Why do the
founders of start-up companies invest their own money in ventures
with high risks of failure? Because their willingness to invest in their
own idea signals their belief in its eventual success.

Spence's model also explains many cases of social behavior. Conspicuous consumption is one example. Buying an expensive car, flash-
ing jewelry, and partying at exclusive clubs is a direct way of signaling
to everyone in your social circle that you have plenty of money, while

also indirectly informing them that you are probably intelligent and professionally successful. It is not surprising that this phenomenon is more widespread in Russia and other former Soviet republics than in the West. Ostentatious displays of wealth in Western countries are not the most reliable signals of one's personal skills because the wealth in question could have been accumulated by previous generations and simply inherited. This does not hold true in Russia. A wealthy adult in today's Russia almost certainly acquired his riches by his own efforts. Showing off that wealth is then a way to signal one's talents.

Ostentatious behavior is not limited to wealth and consumerism. Academics generally have little interest in showing off economic wealth (which they usually don't have anyway). They do, however, have incentive to boast about their academic successes. Ostentatious behavior in this context is expressed by bragging about the number of books or articles published or the number of invitations received to lecture at prestigious conferences. The very same psychology (and economics) is at work when clergymen try to show off their spiritual skills by pointing out how many congregants they have and counting up the number of souls they have managed to save.

Ostentatious behavior is not defined by the means used to express flamboyance but by the flamboyance itself. In the eighteenth century several small and eccentric messianic cults emerged in Jewish communities in eastern Europe. These cults competed with each other by showing off the extents to which they were willing to take religious precepts to the most extreme manifestations. One such group stressed humility above all else, as an expression of man's insignificance before the power of heaven. In their native Yiddish language their motto went "Ich bin gur nicht," meaning "I am worthless." During prayers the congregants would each publicly disgrace themselves in turn, again and again repeating how miserably little their physical existence meant.

One day a new congregant, who happened to be a tall and very distinguished looking man, moved in from another town. Having been informed beforehand of what was expected of him, as soon as the prayer service began he flamboyantly fell facedown on the ground, shouting at the top of his voice "I am but an earthworm before God! I am nothing, less than a grain of sand!" A pair of veteran congregants nearby, watching this spectacle, whispered to each other: "Look at this! He joins us this morning and already he thinks he is nothing!"

Many human societies express values with respect to ostentatious and arrogant behavior that can seem at first glance paradoxical: flamboyance is seen as weakness while humility is regarded as strength. The strength of humility stems from the handicap principle: a person who is always humble refrains from showing off his positive characteristics, which would seem to place him at a disadvantage in social competition. But that is exactly where the power of humility comes from. A person who does not show off is signaling that he is so well endowed that he does not need any external displays to be appreciated, or that he is so high up the social scale that he does not need a boost up.

There is a relationship between the interpretations that societies give to humility and arrogance and the frequency with which these traits are expressed. In societies in which arrogance is the norm, humility is regarded as weakness. In societies in which humility is dominant, arrogance is taken to be inconsiderate behavior expressing an exaggerated self-image.

A few years ago, when I visited Oslo for the first time, I discovered how different social attitudes can be between different countries. I mentioned to my hosts that there was a distinct lack of expensive cars parked on the streets, even in the most posh part of the city, to which they replied that although there is a significant wealthy population in Oslo, it can be very difficult to identify who belongs to it. The difference between high-income workers and average-salaried employees is expressed only in their bank accounts. From the consumer perspective the two populations are indistinguishable. This public humility has nothing to do with avoiding the tax man. As pointed out earlier, although Norway has one of the world's highest tax rates, it also has one of the world's lowest tax evasion rates.

Humility in Norway extends much deeper than matters of money and consumption. When I sought to learn more about one of my hosts, a full professor of economics at the University of Oslo and an extremely sensitive and intelligent person, I came across this self-description he had placed on his Web page right next to his photograph (borrowed from a 1984 paper by Tversky and Kahneman)[2]:

> Kjell Arne Brekke was born in August 1960. He is intelligent but unimaginative, compulsive, and generally lifeless. In school he was strong in mathematics but weak in social studies and humanities. He usually pays insufficient attention to clothing, as can be witnessed by the asymmetric shape of his shirt on the picture. He does, however, play jazz for a hobby.

If arrogance is a signal for self-confidence and strength, then humbleness can be appreciated as a form of the handicap principal, as discussed earlier. For people with established reputations, showing off doesn't buy much. For them, humbleness becomes a much more effective signal for strength. Golda Meir, the only woman who served as an Israeli Prime minister, was known for her political incorrectness. In the early 1970s she received an important US diplomat for talks in Jerusalem. Following the man's speech some of her advisers witnessed her whispering into his ear: "You shouldn't be so humble; you are not so great."

19

OVERCONFIDENCE AND RISK

The "It Can't Happen to Me" Syndrome

KENNETH ARROW IS WIDELY ACKNOWLEDGED TO BE ONE OF THE founding fathers of modern economic theory. Over a period of a decade he and I codirected the Jerusalem Summer School for Economics, which annually attracts leading researchers and doctoral students from around the world. I was thus privileged with the opportunity to engage in a very large number of deep and lengthy conversations with him, many of them on the philosophy of decision making. In one of those conversations he related a story that he heard at a lecture by the statistician and operations researcher, Merrill M. Flood.

At one point during the long fight against Japan, Flood's research unit was tasked with suggesting solutions to a dilemma. The US military attached great strategic importance to conquering the Pacific island of Saipan about 2,000 miles from Tokyo, held at the time by the Japanese. The island was of enormous strategic importance for the United States because it would enable the construction of a forward refueling base for bombers on their way to attacking targets on the Japanese mainland. The direct conquest of the island was to be accomplished by landing Marine invasion forces after a planned massive aerial bombardment of the island's entrenched Japanese units by elite air squadrons.

Operational planners estimated that an immense amount of ordnance would be required for the aerial bombardment to achieve its aims. Dropping that quantity of explosives would require, in turn, that every pilot conduct several bombing sorties, taking off from and returning to an airfield quite distant from the island. Each such sortie exposed the pilots to significant risks from antiaircraft gunners and Japanese fighter planes. In addition, it was clear that the more bombs loaded onto a plane, the more effective each sortie would be. But adding bombs also increased the risks to the pilots. The sheer weight of the bombs, along with the weight of the fuel needed to get to the target and back, limited the plane's maneuverability in the face of enemy fire.

Army Air Force staffers working jointly with Flood's unit worked out an exact relationship between the weight borne by a plane and the risk to the pilot. Flood's unit was charged with mathematically calculating the optimal way of getting the requested amount of ordnance dropped on enemy forces while minimizing the expected number of pilot fatalities. The main dilemma was whether to conduct many low-risk sorties or a small number of high-risk sorties.

After several days of brainstorming, the unit concluded that there was one optimal solution that would minimize the expected number of total pilot fatalities while attaining the operational goals of the mission. All the researchers working on the problem were unanimous in agreement on the proposed solution, which was as follows: a lottery would be conducted among the pilots participating in the mission, selecting one quarter of them. Each of those lottery-selected pilots would then set out on one and only one bombing sortie, with his plane loaded as heavily as possible with bombs. With each plane maximally loaded with bombs, the entire mission could be accomplished by only a quarter of the pilots. The other three fourths would be relieved of any duty in the bombing mission. To enable the planes to get off the ground with that many bombs on board, however, the amount of fuel that each plane flying on the mission would carry would be sufficient for only a one-way flight to the bombing target.

In other words: under the suggested plan one quarter of the pilots, chosen by lottery, were to be sent to their deaths, because flying over enemy territory without sufficient fuel to return to base is a suicide mission. In contrast, the other three fourths of the pilots would bear zero risk of death, because they would not be flying at all.

The calculations actually showed that this plan yielded the lowest overall risk to the flight squadron. Selecting those who would die with certainty by lottery meant that each pilot had a 75 percent of surviving. Under any other proposed solution studied by the researchers, the individual expected survival rate of each pilot was significantly lower.

The pilots, however, were unanimous in absolutely and categorically rejecting this plan. They preferred to divide the bombs equally among themselves, flying many more sorties and taking their chances against enemy fire instead of submitting to a lottery that would determine who would live and who would die. Fortunately, the whole discussion ended when the Marines captured the island of Iwo Jima, only about 600 miles from Tokyo, so that the trade-off between fuel and bombs became much less significant. When I tell my students this story, they often react by claiming that the research unit's proposed solution was immoral because it was inequitable. But they are wrong. It did attain the stated aim of saving the greatest number of lives under the expected conditions in a very equitable manner, because each pilot had an equal chance of being selected for a suicide mission under the lottery. In fact, the solution preferred by the pilots was less equitable because it did not give each pilot an equal opportunity for survival: the less-skilled pilots, or those who were unfortunate enough to have suffered a bad night of sleep prior to a bombing run, carried greater risks than the other pilots.

The explanation I prefer for the pilots' refusal to accept the researchers' solution is related to a phenomenon that has been much studied in the psychology and economics literature: overconfidence.

Most of us, most of the time, fool ourselves into believing that we have greater capabilities than we actually do in reality. This is the famous "it won't happen to me" syndrome, with which we are familiar each time we hear about someone else's failure. If you fail to recognize it in yourself, conduct the following simple experiment with the participation of a group of friends or coworkers. Choose a skill that some members of the group are better at than others, such as driving or cooking. Ask the group members to rate their own abilities using one question: do you believe that most of the people in the group are better or worse than you at this particular skill? If possible, ask this question with respect to more than one skill.

After you have gathered the responses, you may be surprised to discover that a vast majority of the respondents, if not all of them, will claim that they belong in the upper half of the group in their abilities (meaning that each believes he or she is better than most of the group). In this situation, some members of the group must be exhibiting overconfidence. It is impossible by definition for a vast majority of the group to be in the top half.

The Second World War pilots in the story apparently also experienced overconfidence. Each pilot believed that his personal flying skills were sufficiently better than those of the others in the squadron to grant him a better than 75 percent chance of survival against enemy forces. Even though the experts had calculated that the maneuverability of the pilots was severely limited by the sheer weight of the bombs being carried and therefore the success or failure of each sortie was almost entirely determined by chance, not skill, each pilot had a gut feeling that "it won't happen to me." That is why they preferred the illusion of controlling their fate to yielding their fate to the outcome of a lottery over which they had no control at all.

Had the pilots gone on the bombing mission using their preferred method of sending all of them out to face the enemy, I believe the extent to which they were mistaken in overestimating their own skills, and how correct the experts were, would have been revealed. Luckily for all involved, the mission was canceled only hours before it was scheduled, after an alternative solution for refueling American bombers on the way to Japan was discovered.

An interesting study of overconfidence and its effects was conducted in 2000 by University of California researchers Terry Odean and Brad Barber.[1] The researchers studied the actions of stock market investors over a period of several years, focusing on specific decisions along the lines of selling shares of stock A and using the proceeds to buy shares of stock B at the same price. An investor should rationally undertake such a transaction only if he or she predicts that the performance of stock B will outpace that of stock A. In fact, Odean and Barber's data showed that on average each such transaction led to a loss of 3 percent. In other words, not only did investors not see an average profit in their stock portfolios by buying and selling, they actually lost money. Taking into account transaction fees and other overhead costs, the cumulative losses were even greater. The conclusion of the study was that overconfidence was leading to high trading levels and subsequent poor portfolio performances (hence "trading is hazardous to your wealth"). This is one reason many investment advisers recommend investing in index-linked funds instead of individual stocks, while avoiding investment managers who may be susceptible to overconfidence in their abilities to predict the future values of stocks.

Several years ago, five professional investment managers participated in an investment competition for an Israeli newspaper. Each investor was given a large virtual sum of money that could be used for trading over a six-month period. In addition to the five human investors, a sixth competitor, whom the newspaper called "the monkey," was involved. The monkey was actually a computer algorithm that randomly selected stocks for investment at the start of the competition, with that random stock portfolio subsequently held constant for the entire time period.

After half a year had passed, the returns of the portfolios of all the competitors were ranked from best to worst. The monkey came in second place, providing better returns than four human professional investment managers. This is undoubtedly an embarrassing outcome for those who consider themselves experts in the stock market, earning immense salaries for their selection of investment portfolios. The monkey's success apparently was achieved largely by what it avoided doing—frequent buying and selling of shares.

We are not born overconfident. It is acquired by learning. When we make decisions in situations of uncertainty, we need to estimate the chances of each possible outcome. For example, the probability we attach to a stock gaining value or losing value affects our decision of whether or not to buy shares of it. Our estimate of the probability that it will rain tomorrow affects our decision of whether or not to take an umbrella, and the risk of a major earthquake occurring will affect a decision on whether or not to buy earthquake insurance.

Over the course of our lives we get indications that are supposed to enable us to update the probabilities we ascribe to certain events. By taking the past into account, the updated probabilities either strengthen or weaken our beliefs that these events will occur. For example, suppose that there are two small urns in the next room, each containing one hundred coins. In one urn there are fifty gold coins mixed in with fifty copper coins, while in the other urn there are seventy-five gold coins and twenty-five copper coins. If someone in the other room were to choose one of these urns at random by a coin toss and bring it to you, and you were asked what is the probability that you got the preferable urn (the one with seventy-five gold coins), you would probably correctly answer 50 percent.

Imagine, however, that you could sample the urn you were given by randomly pulling one coin out of it, taking a look at it, and placing it back. If you pulled out a gold coin, would you still stick to the 50 percent probability estimate that you received the better urn? Of course not. You have just gotten an indication (but only an indication, not a proof) that this is the better urn. You would update your belief accordingly.

Bayes's Rule (named after an eighteenth-century mathematician named Thomas Bayes) is a precise mathematical formula for conducting these sorts of probability updates after receiving new information. In this example, Bayes's Rule determines that sampling a coin at random and seeing that it is a gold coin updates the probability that you received the better urn to 60 percent. If you then pull out another coin randomly and place it back, and discover that it is a gold coin too, that would trigger yet another probability update, increasing the estimate. But every time you pull out a copper coin is a bad indication, lowering the estimate that you've received the better urn. If the coins in the urns have been properly mixed, after sufficiently many such samplings the probability estimate will be either very close to 100 percent or very close to 0 percent. In either case, you will be very close to certain knowledge of which urn you've got before you.

The reader at this point may be asking what all this has to do with self-confidence. Here is the connection: we never know with certainty whether we are better than average at a particular skill or worse than average. From this perspective, our knowledge of our own abilities is similar to our knowledge of which urn we are drawing our coins from. We do receive daily indications of our abilities. These indications parallel the sampling of coins in the example. Every time we cook, for example, we get an indication of how good we are at cooking. If we burned the scrambled eggs we were preparing for our spouse this morning, we get a negative indication (similar to sampling a copper coin from the urn), and we should accordingly reduce our estimation that we are better than average at cooking. If we have guests over for a multiple course meal that we have cooked, and the guests clean off their plates to the very last drop of sauce, we get a positive indication (parallel to sampling a gold coin). The same reasoning holds regarding our capability at taking good photographs, choosing good financial investments, or making strong social connections—in each skill in life we rate our abilities using indications we get, with each indication leading us to update the probability that we are better or worse than average. We don't, of course, normally make formal use of Bayes's Rule in everyday life. We use our memories and intuitions. Each indication is stored in our mind's memories, changing our beliefs a bit. In many cases our intuitive updating is quite close to what Bayes's Rule indicates we should do.

If you have been following so far, you must be asking yourself—so what's wrong? Why are you accusing us of overconfidence? The answer is that we update our beliefs quite well as long as we are not assessing ourselves. When we come to assess ourselves, we start fudging the arithmetic in our favor without even noticing we are doing it. Bayes's Rule demands giving strictly equal weights to positive and negative indications. But our cognitive and emotional systems refuse to do this. Consider again the example of self-assessment of cooking skills. Most of us give greater weight to our successes in cooking than to our failures, stressing the most successful meals we cooked and forgetting the ones we burned.

Uri Gneezy, Muriel Niederle, and Aldo Rustichini coauthored a paper on this phenomenon in a very convincing laboratory experiment in which students were asked to solve relatively simple riddles while reassessing their abilities at doing so from one riddle to the next.[2] The students systematically refused to take their failures into account as much as their successes. As a result they tended to overestimate their abilities in solving riddles. In the experiment, in addition to rating their abilities, the students also bet on their chances of solving the next riddle. Their overconfidence caused them on average to lose money in those bets.

Researchers have only recently started to unveil why we overestimate ourselves, but it is reasonable to suppose that our emotions are particularly active in causing the phenomenon. Our reactions to our successes and failures are mainly emotional, accompanied by feelings of happiness or disappointment. Our emotional reactions may be affected by our ability to remember the indications we get over the course of our lives. We have selective memories that stress the positive events in our lives and blur the negative memories.

Learning from past experiences is undoubtedly an important element in ensuring survival. Imagine what would have happened if prehistoric men were unable to learn from their failures in hunting, arriving time and again at the same clearing in the forest, only to see the prey they are stalking always slip away from them into the woods. Given that, why hasn't evolution given us the ability to learn about our self-worth efficiently, protecting us from overconfidence?

The answer is that along with the damage it causes, our bias toward overconfidence also bring with it advantages, in fact several important advantages. First of all, self-confidence plays a role similar to the peacock's tail in raising our "market value" in many social interactions, including the most important interaction from the evolutionary perspective—that relating to reproduction.

Overconfidence also gives individuals advantages in the competition over resources and territory, because expressions of self-confidence can intimidate rivals. Just as at equilibrium an individual's emotional state can effectively impress others only if it is authentic, faked self-confidence is not as effective as the real thing. If you want to convince others that you have strong abilities, you had better really believe it yourself.

The third advantage of overconfidence is that it can encourage optimism, even a bit of overoptimism. Optimism sparks action, and action is good for survival, hence optimism is good for survival. Imagine, again, two prehistoric hunters, one somewhat optimistic, the other somewhat pessimistic. The optimistic hunter wakes up in the morning and eagerly grabs his hunting implements, believing that this is the day on which he will bring down the fattest buffalo on the plains. The pessimist, in contrast, curls up deeper under his deerskin blanket in his cave instead of getting up in the morning, while mumbling about what an idiot his optimistic mate is: "Doesn't that misguided optimist realize that he can hop from one hill to another all day long, waving his sharpened spear, and still come home at sundown with nothing to show for all of his efforts?" Guess which one of these two hunters has better chances of bringing home a buffalo.

An extensive psychiatric study published in 1989 compared probabilistic assessments expressed by psychologically healthy people with those suffering from clinical depression.[3] Individuals in both categories were asked to assess the chances that they would experience negative events such as falling ill, being injured in accidents, losing their jobs, and so forth. In addition, they were also asked to assess the chances that they would experience positive events such as finding a spouse, winning money in the lottery, and so forth.

When the researchers compared the answers given by both groups with the true objective probabilities of each event, they found that the clinically depressed, who were also quite pessimistic, were much more accurate than their healthy counterparts in assessing the probabilities of both positive and negative events. Depression, it turns out, makes you much more realistic. Nevertheless, it is difficult to conclude from this that depressive realism confers great survival advantages. Quite the opposite. It is the nonrealistic illusions that healthier individuals live with that make daily living easier and give us better chances at survival—assuming that the rosy illusions do not get too far away from reality. An overdose of self-confidence can be fatal.

Niederle and Vesterlund conducted another interesting study relating to self-confidence, this time comparing men and women.[4] In contrast to popular opinion, men do not have greater self-confidence than women. Both sexes share equally in the bias toward overconfidence. But a significant difference was noted in the ways that men versus women update their self-assessments after receiving indications. Men, in general, are better at updating the probabilities regarding their own abilities. They give sufficient weight to both positive and negative indicators and more readily change their initial assessments. Women, by contrast, are more stable in their self-assessments (whether those are low or high). Successes and failures have less of an effect on their self-assessments.

It is possible that these sex-related differences have an evolutionary basis, but even if the evolutionary basis for differences in self-assessments is marginal, social interactions tend to amplify it. In the market for mates, each individual has an interest in broadcasting the characteristics that are most closely associated with that individual's gender (that is, women wish to stress their femininity while men show off their masculinity). Sexual attraction works in most human beings quite similarly to the way it works in other highly developed animals—each individual seeks a mate with as many characteristics of the opposite sex as possible.

There is a story about the legendary investor Warren Buffet receiving a telephone call from his wife one day while he was driving on Route 1 near Boston. "Warren, drive carefully," said his wife. "I just heard on the radio that there is some idiot driving against the flow of traffic on Route 1."

"My dear," replied Buffet, "I wish it were only one idiot. I see dozens of cars doing so!"

In this joke, Buffet exhibited more than supreme overconfidence—he also showed off his nonconformism and his refusal to bow down to social conventions. As we will see in the next chapter, however, despite our tendency toward overconfidence, the truth is that we usually behave in very conformist ways.

20

THE VOICE IS HERD

On the Sources of Herd Behavior

OCCASIONALLY WHEN I AM AT DINNER WITH FRIENDS AT A FINE RESTAURANT I notice a peculiar behavioral phenomenon. After we have all intently studied every item on the menu from start to end and even discussed what we might want to order and what we want to avoid, the moment of decision arrives. If it is my rotten luck to be the first person whom the waiter approaches, I try to make a brave choice, expressed in a determined manner.

When the next person in turn is asked by the waiter what she wishes to order, I look at her with compassion. But when the third person orders, I start raising an eyebrow and by the time it is the fourth person's turn, I really start to sweat. After that I don't even bother listening to what the rest of the party are ordering—I just know I have made a fatal mistake. There is nothing left to do but wait anxiously until the waiter is done taking the orders and walks away. At that point I weakly apologize to my friends and run to the kitchen to change my order.

If you go through even a pale version of this, you are not alone.

Our overconfidence evaporates the minute we are asked to make a decision in parallel with others or after others around us are deciding on the same question. That is when we are most susceptible to conformism, to copying others and dismissing our own opinions much too quickly in the face of majority opinion.

Our tendencies toward this type of conformism do not necessarily contradict our biases toward self-confidence. Self-confidence relates to our subjective judgment of our own abilities, while our tendency toward conformism is often due to mistaken information processing. Sometimes it stems from a fear of being perceived as peculiar.

The herd phenomenon has important implications in a wide range of social situations. Hundreds of studies have been composed on the subject in economics, finance, and psychology. To some extent herd behavior is responsible for many financial market crashes, along with the bubbles preceding them. It is also the reason many erroneous stigmas spread so easily (such as, "if I see that none of my acquaintances have hired employees with disabilities, then it is probably better for me also to avoid hiring such a person"). It is responsible for the sort of homogeneity in thought and behavior that depresses creativity and renewal in societies. But the worst effect of herd behavior is that it can cause an immense number of individuals to make wrong decisions in a dynamic process, with each person influencing others around him in a misguided way, despite good intentions.

Imagine yourself on holiday in Malaga, Spain, looking for a good place to eat lunch. After an hour of exhaustive search, hungry and tired, you decide that you will enter the next restaurant you pass, no matter what. After a minute, you find yourself looking at two adjacent restaurants: one so crowded that there is barely an empty table to be seen, the other as empty as a ghost town. It is not difficult to guess which restaurant you will choose. Researchers debate whether your decision to enter the crowded restaurant stems from efficient information processing or alternatively whether a misguided herd that knows nothing about restaurants has dragged you into making the wrong decision.

We will use this example to illustrate how herd behavior can occur even if every single person is acting fully rationally, meaning that everyone satisfies the following conditions:

1. Individuals have their own sources of information that they use to arrive at correct decisions.
2. Each individual perfectly well understands how to use probabilistic models and is not limited in calculation ability.
3. Individuals seek to maximize their own utility.

It is entirely possible that even under these perfect conditions of rationality, herd behavior can lead everyone to the worse restaurant.

Let's call one of the restaurants Salvador's and the other one El Torero. Let's further suppose that Salvador's is a better restaurant than El Torero. Suppose now that there are one hundred tourists on a particular day trying to decide whether to eat at Salvador's or El Torero. With these assumptions, I will now describe a process that will lead all one hundred tourists to El Torero in an entirely rational and well-calculated manner.

Suppose that prior to arriving in Malaga, each tourist looks up some information on the city's restaurants. That information is not enough to determine decisively which restaurant of the two is the better one, but let's suppose that each tourist slightly prefers Salvador's. This can happen, for example, if each tourist ascribes a 51 percent chance to Salvador's being a better restaurant and only a 49 percent chance to El Torero's being the better one (which could happen, for example, if a popular tourist guide book notes that Salvador once ranked higher in the Michelin restaurant ranking).

On arriving in Malaga, the tourists receive another indication of the relative qualities of the restaurants (such as an email from a friend, a Web site ranking, or a recommendation from a hotel clerk). It is reasonable to assume that since Salvador's is objectively better, there will be more positive indications for Salvador's than El Torero. But there is some random element to these recommendations. A tourist could, for example, have received an email from a friend who happened to wander in the past into El Torero and liked the food served there (it is not a bad restaurant, after all, just not as good as Salvador's).

Based on the new information received, each tourist now updates his probabilistic assessment of the relative qualities of the two restaurants using Bayes's Rule (as described in the previous chapter). Recall that we assumed that all of the tourists are not only rational, they are experts in probability theory. Suppose further that all the indications are sufficiently strong that after this updating process each tourist has a high level of confidence that he knows which restaurant is truly the better one. Given the rationality that everyone exhibits, a tourist who receives only one positive indication for one restaurant but two positive indications for the other restaurant updates his probabilities in such a way that he ascribes higher probability to the restaurant with two positive indications being the better one.

✳ ✳ ✳

AND NOW, ON TO THE MAIN COURSE. IMAGINE ALL ONE HUNDRED tourists standing in a line at 11:59 a.m., waiting for the two restaurants to open their doors to the lunchtime crowd at noon. Each tourist has received one indication favoring one restaurant or another, with the two tourists at the head of the line having received positive indications for El Torero (remember again that some of the tourists have received recommendations for El Torero, and it is not surprising that the two who happen to be at the front of the line may be among them).

At noon the front doors of the restaurants swing open. The waiters in each of the so-far empty restaurants wait in anticipation for the lunchtime crowd to enter. Every successive tourist in the queue decides, in turn and entirely rationally, where he will eat. The tourist at the head of the line, based on the positive indication for El Torero that he received up to that moment, naturally chooses El Torero. The second tourist, who has also received a positive indication for El Torero, does the same.

What about the third tourist? Let's suppose that prior to noon she has received an indication that Salvador's restaurant is the better one. However, she has just seen the two people ahead of her in the queue choose El Torero. She thus surmises that they each received positive indications for El Torero (which clearly differ from the indication she received). She can now take this new information into account in making her decision: she knows that there are two indications for El Torero (based on the choices of the two people in front of her in the queue) and only one indication for Salvador's, which she previously received. That makes a majority of two to one in favor of El Torero. The third tourist thus promptly enters El Torero for lunch, overriding the indication that she had personally previously received. In other words, the third tourist will choose El Torero regardless of the signal that she herself got.

The fourth tourist is in a situation similar to the one that the third tourist was in. He knows that he cannot really learn anything from the behavior of the third tourist, who chose El Torero independently of the indication that she received. But he does know that the first two tourists did receive positive indications for El Torero. From his perspective, that constitutes a majority of positive indications for El Torero over Salvador's, and therefore he, too, goes directly to El Torero for lunch.

It should be clear to anyone by now how this interesting lunchtime crowd of tourists is going to behave. Each and every tourist, based on the choices of the first two tourists (the choices of the rest are irrelevant, since they are basing their choices on those of the first two) will choose El Torero over Salvador's, using the same reasoning as the third tourist. And thus the poor owner of Salvador's, who really has worked hard to produce a meal superior to that of El Torero, will spend the entire afternoon in his empty restaurant, sadly watching his rival El Torero filled to the rafters with a herd consisting of every single tourist in town.

The story I have just told is based on a mathematical model appearing in a paper published in 1992 by three professors of finance at UCLA.[1] The authors of that paper claimed that herd behavior typically occurs as a result of the most rigorously rational thinking, as in their model, and not because of psychological biases such as conformity, lack of self-confidence, and so on. It is quite an ingenious observation (if slightly contrived) that perfect rationality can still lead to herd behavior. But is this really the way herd behavior actually happens?

It was precisely to answer this question that three colleagues (from the Max Planck Institute in Germany, the University of Paris, and the University of Aberdeen) and I conducted a research study that featured a laboratory experiment in which we induced herd behavior.[2] In our experiment, subjects were not asked to choose between restaurants; we based the experiment instead on the urns described in the previous chapter.

Two urns were filled with balls, one hundred balls in each urn. The first urn contained fifty red balls and fifty black balls. The second urn was filled with twenty-five red balls and seventy-five black balls. The subjects of the experiment were informed that one of these two urns would be selected, with the first (50–50) urn selected 51 percent of the time and the second (75–25) urn selected 49 percent of the time. They were also told that they would be rewarded monetarily for correctly guessing which urn was actually selected. Each subject in turn was given one opportunity to secretly remove a ball at random from the urn, check its color, and then place the ball back in the urn. After doing so, he or she was to announce publicly, in front of all the other subjects in the experiment, his or her guess as to which urn was selected (note that this public announcement parallels choosing one of the restaurants in the above story, with correctly guessing the identity of the urn parallel to correctly choosing the better restaurant).

As expected, we managed to create significant herd behavior in the lab. The herd usually began to form after three or four identical guesses had been made out of nine, that is, after the first three participants had publicly announced the same guess, the six others in each experimental round made the same guess, independent of which color ball they pulled out of the urn.

In the second stage of the experiment we carefully tested whether the explanation for the occurrence of herd behavior suggested by the three UCLA professors held up to scrutiny. Note that their explanation depends crucially on the assumption that after the first two tourists have made identical choices all the rest of the herd follows their example, but everyone else is doing so knowing that they can only learn something from the behavior of that first pair, not from the behavior of all the others. In other words, when the one hundredth tourist sees the ninety-nine tourists preceding him entering El Torero, his level of confidence in the choice of El Torero as the better restaurant is identical to the level of confidence of the third tourist who has only seen two tourists before him choose that restaurant. Both are basing their decisions solely on the decisions of the first two tourists.

That seemed unrealistic to us. If that were true, it would mean that if we were to give the one hundredth tourist a slightly better indication than that given to the first two tourists, he would choose based solely on the indications he received personally, even after seeing ninety-eight tourists ahead of him choosing differently (since he is supposed to ignore the behavior of everyone but the first two tourists in the queue). We did the equivalent thing in our experiment in order to test these assumptions. Selected subjects in our experiment, at various points in time during the development of the herd behavior, were given significantly better indications than others regarding which urn had been selected.

If the UCLA professors' explanation were correct, these subjects should have always followed the indications they received, independently of the intensity of herd behavior that they were witnessing. But that is not what happened. When the herd behavior was just beginning to develop, and only a small number of subjects had made identical guesses, the subjects given private extra indications did indeed follow those indications to a greater extent than they followed the herd. But after the herd behavior had gathered strong momentum, they ignored their private indications and joined the crowd, as we had expected. Our conclusion was that the UCLA explanation did not stand up to close scrutiny. Herd behavior is much more stable and less fragile than their model would indicate, and it cannot be explained within a purely materially rational framework.

It is unreasonable to expect there to be one dominant explanation for herd behavior. The context in which the herd phenomenon develops is relevant. Even in phenomena such as real estate bubbles or stock market crashes there are several forces at work. When the stock market enters a downward spiral, we usually rush to sell our shares for at least two reasons: first because falling stock prices may be an indication that market fundamentals have taken a downturn and our expectations of seeing share profits correspondingly fall. But even if we are perfectly assured that the fall in prices is due solely to irrational panic while market fundamentals remain strong and stable, we are perfectly justified in selling our stock holdings as quickly as possible. With everyone else selling, the longer we hold on to our stocks, the less they are worth, minute by minute. In other words, it is quite possible that everyone rationally knows that there is no fundamentally sound reason at all to sell stocks and flee the market, yet everyone does exactly that because of the expectation that everyone else is going to do that.

Most financial crises, in fact, are caused by such self-confirming expectations. It is precisely in such situations that government intervention can be used most effectively for rebuilding trust and cooperation, reducing the fears driving investors to flee the market. This is why many governments extend deposit insurance for bank accounts. Without it, bank runs would be far too common.

In contrast, in many situations herd behavior develops because people feel a desire to join certain groups. The rapidity with which clothing fashions, artistic styles, and even ideologies spread in societies are examples of this phenomenon. There is no role for information and updating probabilities here, only a desire on the part of some individuals to be identified with other individuals. Many instances of herd behavior arise from the types of collective emotions discussed in an earlier chapter.

There is another phenomenon studied in the economics literature that is not considered to be herd behavior but is definitely related to it: peer effects. This occurs in situations in which peers (work colleagues, fellow students, and so forth) tend to copy each other's behaviors. Bruce Sacerdote, a Dartmouth College economist, published a study in 2001 of how peers influence the extent to which students invest time and effort in university studies.[3] Students of various backgrounds, majoring in different subjects, were assigned to student dorms, two to a room. The students had no input or influence on these rooming assignments, which were effected entirely randomly. Despite this, by the end of the academic year dorm-mates exhibited strong degrees of correlations in the grades they received. The study's conclusion was that these correlations were formed by mutual influences between dorm-mates. A student who conscientiously devoted time for studies apparently influenced his or her dorm-mate.

Similar phenomena have been noted among work colleagues in several research studies. But workers have a positive incentive in getting their colleagues to invest efforts in hard work (because the harder their colleagues work, the more successful the workplace will be, to the advantage of all the workers). It is more difficult to explain why peer effects, with respect to investing efforts in studies, should appear among students of diverse backgrounds and majoring in different subjects. One possible explanation is simply a human tendency to copy the behavior of others, but the phenomenon might also stem from competitiveness.

The simplest and most general explanation for the diverse varieties of herd phenomenon, in fact, goes back to the distinction between rule rationality and act rationality that was described earlier in this book. Correctly processing information is a very difficult task to accomplish. Experts often fail at it. To illustrate just how difficult it is to use correct probabilistic reasoning for making decisions, consider the following three stories, taken from scientific journals:

1. *Nature Neuroscience,* one of the leading journals in the field of brain studies, published a paper in 2011 looking into common mistakes in probability calculations made

by neuroscientists. The authors reviewed 513 papers published in the foremost brain studies journals over a period of
two years.[4] They found that in 157 papers in which errors
in probability could have been made, half contained such
errors, compromising the conclusions that they reached.

2. One of the most impressive experiments conducted by
   Daniel Kahneman, a Nobel Prize winner in economics,
   along with his long-time collaborator Amos Tversky, dealt
   with the abilities of physicians to process probabilistic calculations in their decision making.[5] Kahneman and Tversky's
   simple experiment involved as subjects medical interns at
   leading hospitals in the United States. The interns were presented with true data on cancer mortality rates in patients
   in the first five years after their initial diagnoses of cancer,
   based on the types of treatments they received: surgery
   versus radiation treatment. Two separate groups of interns
   were given exactly the same data, but it was expressed in
   different ways. One group was informed what percentage of
   cancer patients died over a five-year period while the other
   group was informed what percentage survived over that
   same period (for example, if one group was told that 60
   percent of patients treated by surgery died within the first
   five years, then the other group was told that 40 percent of
   patients treated by surgery survived the first five years). Obviously, both sets of data were saying exactly the same thing.
   Despite this, the two groups of interns gave very different
   treatment recommendations, depending on how the data
   were presented to them.

3. Maya Bar-Hillel, a student of Daniel Kahneman, conducted
   an interesting experiment, using senior Israeli court judges
   as subjects, to study the extent to which they understood
   principles of probability. Given that the Israeli justice system
   (like those of all Western nations) is based on a standard
   of evidence requiring "proof beyond reasonable doubt,"

Bar-Hillel was interested in ascertaining what the judges regarded as reasonable doubt and whether they correctly apply the standards they are sworn to uphold. To achieve this, she presented the judges with examples of evidence and asked them to decide whether or not the examples satisfied the requirement of providing proof beyond reasonable doubt.

Here is one example of the sort of evidence that Bar-Hillel used in this study, slightly reworded: a motorist asked for a court to review a parking ticket that he was issued when his car was parked at a location with a maximal continuous parking time of one hour. A traffic warden testified that he had twice seen the car parked at the same spot over a period of an hour and a half. In his defense, the motorist claimed that he had parked at that location for three-quarters of an hour, moved the car backward to the spot behind him, and then returned to the same parking spot fifteen minutes later, hence he had not parked continuously in the same location for over an hour.

The traffic warden retorted that in this case he had conducted detailed surveillance of the car's position by recording the positions of the air-pressure valves of each of the car's four tires (assigning one of the four positions: north, south, east, and west) both times he saw the car parked at the same spot. In each case, their positions were identical. The claim that followed from this observation was that it is unreasonable for the car to have been moved and then returned to the same location with the four air-intake valves restored to their exact positions. Most judges tended to agree with this claim. They explained that if this had been observed of only one tire, they would be less inclined to accept the evidence. But if it was observed in all four tires—now that was convincing.

Only a few judges noticed that, in a straight and short move of the car, if the position of the air-intake valve of one tire had been restored to its previous position, then it is almost certain that the same holds true of all four tires. In fact, the probability that the positions of the air-intake valves would return to the same position entirely randomly turns out to be almost 25 percent—making it rather reasonable to suppose that the motorist could indeed have moved the car and later returned to the same spot.

Since we are, in general, unable to make efficient decisions when faced with the need to undertake complex probabilistic calculations, we all tend to use heuristic reasoning instead. The heuristic that supposes that "the majority is right" is a simple one that serves us well in many real-life situations. The herdlike behavior that results is also unfortunate, but it is ultimately an acceptable side effect.

21

TEAM SPIRIT

The Paradox of the Generous Bonuses and the Lazy Workers

WORKPLACE BEHAVIOR IS ONE OF THE MOST IMPORTANT SUBJECTS STUDIED in economics in general and behavioral economics in particular. I myself have devoted a great deal of attention to this topic in recent years.

One elementary reason this subject is of such importance is the fact that the single most expensive factor of production in just about every commercial enterprise is human labor. The amounts of money that corporations and organizations spend on human resources overshadow all their other expenditures. Proper planning of incentive structures at workplaces (one of the main subjects studied in human resources economics) can pay dividends in significant savings of unnecessary expenses but, perhaps even more importantly, can also lead directly to increased production and profits.

A classic case of the bottom-line miracles that a company can bring about if it uses the right incentives happened at Continental Airlines in the years 1992–1997, as detailed in a paper by Marc Knez and

Duncan Simester.[1] In the early 1990s, Continental Airlines suffered a financial crisis so severe that the company registered a $125 million loss in 1992. Internal company audits identified late departures and landings as the main cause of the enormous losses, which continued to balloon out of control, swelling to a $199 million loss in 1993 and a staggering $619 million in 1994. This amount of money hemorrhaging could not, of course, continue indefinitely, and Continental Airlines came perilously close to outright bankruptcy.

Continental executives correctly reasoned that they had to change the incentives they were offering employees if the company were to survive. Ensuring that airline flights adhere to tight time schedules is a "production process" that is only as reliable as its weakest link. A delay in even one of many preflight inspections and preparations that is required prior to every take-off is sufficient to cause significant lateness in getting the plane off the ground. After a long series of meetings analyzing the issue, management decided it would make one last, desperate attempt to save the airline from extinction by implementing what they code-named the Go Forward plan.

One of the central elements of Go Forward was the promise of a $65 bonus to every employee following every month in which the company was ranked among the top five airlines in on-time departures and landings. The plan's effect was dramatic and immediate. Within one year, in 1995, Continental went from a $619 million loss to $224 million in profits.

Profits continued to soar, reaching $385 million in 1997. Interestingly, the monetary incentive aspect of Go Forward, which played a crucial role in saving the company, was a collective incentive, not a personal one. Teamwork was being rewarded, not outstanding individual effort.

So, here is the question that we'll ponder in this chapter: what did the Go Forward initiative get right? As a business manager, how can you emulate this program to get better results?

Teams of workers in places of employment are a very interesting microcosm of social interaction, involving both rationality and emotions. Teams constitute an important element in economic and organizational efforts virtually everywhere around the world. A 1995 survey conducted by Paul Osterman revealed that more than 54 percent of organizations in the United States base their workforce activities on teams. In commercial companies that number was even higher, at 66 percent.

Advanced understanding of the behavior of teams in workplaces requires extensive use of mathematical models and game theory, because without first understanding how we should expect team members to behave under assumptions of rationality and self-interest, we cannot expect to understand how psychological and emotional phenomena contribute to a team's success or failure. I have personally conducted several research efforts in recent years studying team behavior using game theory, more specifically a subfield of game theory called contract theory.

A contract between two or more individuals can be viewed as a game, because each contract defines rules of interaction (which parallel strategies in games) and furthermore specifies the payoffs to each party to the contract as a result of the actions undertaken by the parties within the framework of the contract (parallel to the payoffs in a game, which are determined by the strategies used by the players). This is how game theory enables us to answer questions about contract planning and contract negotiations. Using game theory we can ascertain what is the best contract from the perspective of contracting party A that will also be acceptable to contracting party B.

Research into this subject has expanded extensively in recent years. It now embraces many different research methods, both theoretical and empirical. Data on the subject have been collected based both on real-life observations and laboratory experiments. Some of the observations have been quite surprising, because they contradict our most basic intuitions regarding teamwork.

One surprising result reveals that the relationship between monetary incentives and work motivation is far from being as straightforward as is usually imagined. A paper I published in 2009 showed that, in the context of workplace teams, personal monetary incentives can have the effect of reducing work motivation.[2] This has nothing to do with psychological effects, nor is it related to the distinction between internal motivation and monetary motivation (which was discussed in a previous chapter, as detailed in a research paper by Gneezy and Rustichini). It can occur in teams even when each individual is selfishly concerned solely with personal gain. I will present a small and simple model of teamwork to try to illustrate this paradoxical phenomenon and its practical implications.

Presenting all the details of this model will take several pages and will include logical arguments that some readers may find challenging. Readers who wish to do so may skip this presentation, as the rest of the chapter is self-contained.

Imagine yourself as the owner of a software company with two employees: one, named Mr. D, responsible for development, and the other named Mr. M, responsible for marketing. The software produced by the company can be profitable only if both development and marketing are conducted successfully. Each employee can choose one of two possible behaviors: investing a lot of effort at his job or investing very little effort. An employee who invests a lot of effort is guaranteed to succeed at his task, but if he invests little effort his chances of success are only 50 percent.

Software production and marketing are conducted in two consecutive stages: first Mr. D develops the software and then Mr. M markets it. The two employees are in different situations not only because the bulk of their work occurs at different time periods but also because Mr. M can see whether or not Mr. D is investing a lot or little effort, while Mr. D has no way of knowing whether Mr. M will work hard or not. You, as the owner, have the responsibility of fashioning an incentive system for your employees that will maximize the expectations that they will work hard. Unfortunately you have no way to monitor the true extent of the effort that each employee is investing in the job. The only thing you can measure with certainty is whether or not your company is seeing a profit (i.e., whether both development and marketing have been conducted successfully or at least one of them has failed). The incentive system you are considering offers both employees a bonus only if the project is successful and the company sees a profit.

To complete the model we need one more important detail: the extent to which an employee suffers from investing a great deal of effort at work. After all, if the employees enjoy putting in extra effort, there is no need to give them an incentive by offering a bonus. We will suppose that the suffering engendered by the effort required is equivalent to $1,000. This, however, does not mean that $1,000 is a sufficient amount to offer as compensation to each employee. The reason is simple: if $1,000 is the offered bonus, to be paid only if the project succeeds, then Mr. D will have no incentive to make an effort, because by making such an effort he will "lose" $1,000 worth of hard work while risking getting nothing if Mr. M does not contribute his efforts to making the project a success. If Mr. M does not invest in making an effort, the probability that the project will succeed will fall to 50 percent, in which case Mr. D will get compensation for the $1,000 worth of work he put into the project with a probability of only 50 percent.

Keep in mind that you, the owner of the company, have no way to know after the project has been completed which of the employees, Mr. D or Mr. M, has invested effort—both, one of them, or neither of them may have done so. The only thing you know is whether or not it has succeeded. How much, then, should you offer each employee as a bonus?

Consider the following sample bonus structure: Mr. D is offered a bonus of $1,400 and Mr. M is offered a bonus of $2,010 (with the bonuses paid only if the project succeeds). Let's try to work out the rational considerations that the employees will take into account, assuming that they care only about their own personal welfare. Start with Mr. M, and suppose that Mr. M has observed Mr. D investing a lot of effort in the development phase of the project. At that point, if Mr. M decides not to make a strong marketing effort then the project will succeed with 50 percent probability. That means that he will receive $2,010 with 50 percent probability, which is equivalent to getting $1,005 in certainty (in economic terms this is the called the "expected payoff"). If, in contrast, he decides to make a strong marketing effort, he will suffer $1,000 worth of hard work but receive a $2,010 bonus with 100 percent certainty. On balance, he receives a total of $1,010, which is a higher sum than the expected payoff of $1,005 that he gets without putting in an effort. In this case Mr. M is better off investing effort in marketing (even though the payoff difference between making an effort and not doing so is only $5).

If Mr. M observes Mr. D not investing any effort in development, then he clearly has no incentive at all to make an effort himself, because under that condition the probability that he will receive a bonus drops to 50 percent even before he gets started. If he also makes no serious effort, the probability that the project will succeed now drops again, to 25 percent. Thus Mr. M's expected payoff is only $502.50 (equal to one-fourth of $2,010). An investment of effort on his part is worth only $5 to him (an expected bonus of $1,005 minus the cost of $1,000 in the hard work he put in).

Next, let's analyze the considerations of Mr. D, whose role in the software development/marketing process comes before that of Mr. M. If Mr. D invests a lot of effort in the development phase, he knows that Mr. M will also work hard (using the reasoning described in the previous two paragraphs), and the project will therefore definitely succeed. In that case, Mr. D will pay a cost of $1,000 worth of hard work but receive in compensation $1,400 in a guaranteed bonus. On balance he will receive a net payoff of $400.

Suppose instead that Mr. D decides not to invest any extra effort in development. Then Mr. M certainly will not invest any extra effort in marketing (as explained above). In that case the probability of the success of the project plummets to 25 percent (which is the probability that the project succeeds if neither employee invests any hard work into it). That means that Mr. D will receive the $1,400 bonus with a probability of 25 percent, translating into an expected payoff of $350. The conclusion of our analysis is that both Mr. D and Mr. M will decide to invest a good deal of hard work in the project, with the two of them earning a combined bonus of $3,410.

Now suppose that you, as the owner of the company, decide to offer the employees significantly higher bonuses, either out of empathy for the work they will have to do or because you believe that doing so will increase their incentives to work hard for your company. The new bonus offers are $1,900 to Mr. D and $4,020 to Mr. M for the successful completion of the project. Just as before, if Mr. M observes Mr. D investing extra effort, then it is worthwhile for him to do the same (because he will now receive a net payoff of $3,020 if he works hard and an expected payoff of $2,010 if he does not).

But consider what will happen if Mr. M observes Mr. D being lazy during the development phase. In that case Mr. M has a 50 percent probability of receiving his bonus if he works hard and only a 25 percent probability if he is also lazy in his marketing efforts. The first possibility grants Mr. M $1,010 (after taking into account the loss of $1,000 he experiences because of the extra work he puts in) and the second possibility yields him an expected payoff of only $1,005 (which is $4,020 divided by four). The conclusion is that in this case Mr. M always has an incentive to work hard in marketing, whether or not Mr. D has worked hard in the development. We see here that the attractive bonus offered to Mr. M does indeed boost his incentive to invest effort in his work. The stakes are simply too high for him, and he would prefer to exert effort even if Mr. D decides to shirk.

But the fact that Mr. M will be incentivized to work hard no matter what Mr. D chooses to do now changes Mr. D's incentives. Previously, we concluded that Mr. D would have an incentive to invest extra effort on the development because he understood that if he were instead to be lazy, then Mr. M would observe that and definitely choose not to invest any extra effort himself. But under the new conditions Mr. M will choose to invest extra effort no matter what Mr. D does. What, then, is the best choice of action for Mr. D? If he works hard, the project will definitely succeed and he will receive a $1,900 bonus—but taking into account the price of $1,000 of hard work he will have to put in, the net payoff to Mr. D is only $900. Alternatively, should he choose not to work hard, the project will have a 50 percent probability of succeeding (since Mr. M will work hard no matter what), which translates into an expected bonus of $950—a higher payoff than what he receives if he invests a lot of effort into the project. In other words, Mr. D's incentive to work hard has been reduced because he knows that Mr. M will do so in any event. Increasing the bonuses has actually worsened the situation. Previously both employees had an incentive to invest a lot of effort in their work; now only one of them is incentivized, even though both of them are rational and seek to maximize their material welfare.

This paradox arises from the mutual influences the employees have on one another. One employee's incentives influence the other employee's incentives. Promising the second employee too high a bonus reduces the indirect threat against the first employee, namely that if the first employee avoids hard work, then the second employee will also refrain from investing heavily, thus harming the interests of both of them.

Planning incentive structures can be a difficult task, one that must be conducted carefully. Our intuitions can easily lead us astray, with significantly negative consequences. I have termed this phenomenon, in which increasing bonuses offered to all the employees paradoxically reduces their incentives to work hard, "incentive reversal." Even though the explanation may seem quite specific and technical, the emotional logic suggests itself often in reality. My research indicates that incentive reversal is quite a general phenomenon that can occur in nearly every organizational structure and any sized workforce.

I recently conducted a large lab experiment, along with several colleagues at the Max Planck Institute in Germany, which elicited a clear and strong instance of incentive reversal.[3] The explanation I presented above for the existence of incentive reversal was based on the fact that the compensation an employee received for investing extra effort in his job was higher when the other employee also invested extra effort. In the example of the employees of the software company this characteristic followed from the development and marketing process, in which the weakest link in the chain determined the probability that the project would succeed. In other words, the employees complement one another: the success of the project depends on the success of both the development stage and the marketing stage. What would happen if the employees did not complement each other as described above, and instead were substitutes for each other (which would be the case if, for example, both of them were developers and the success of the project depended on only one of them making a successful effort)? The incentive reversal paradox would not happen in that case. This conclusion follows from both the mathematical model and the data observed in the lab experiment.

Next, consider what would happen if neither employee could observe the amount of effort that the other employee invested in the project. What then would be the best contract to offer them? In an article that I published in 2004 I mathematically proved that in that case the best thing to do is to create a certain discrepancy between the employees' bonuses, even if they are entirely identical in their roles and capabilities.[4] This is because an employee who is offered a lower bonus has added incentive to invest extra effort because he is then certain that the other employee (who stands to receive a larger bonus) will also invest extra effort.

That article received considerable attention. Some claimed that the theoretical advantage of discriminating between employees in the bonuses offered to them would be canceled by the anger against such blatant discrimination that some employees would feel. But two groups of experimenters who independently tested the theory in lab experiments came to the conclusion that a certain amount of moderate discrimination between employees, as described above, does indeed lead to increased work incentives.[5,6] We naturally recoil from inequality, but when it serves our interests, we tend to accept it, even when we are the party who gets the lesser end of the deal.

In the rest of this chapter, we will no longer assume that individuals selfishly care only about their own material benefits and instead consider more realistic work environments in which self-interest, psychology, and emotions all interact. In such environments peer effects play significant roles, adding emotional and social incentives over and above monetary incentives that often improve team performances. Peer effects can cause an employee to avoid investing too much effort in her work if she knows or believes that most of her co-workers are themselves shirking from putting in extra effort, but they can also spur her to work much harder if she believes that the others are doing the same. Following are three examples of this, taken from the economics literature.

Italy, more than any other European Union country, is characterized by extreme cultural gaps between different geographical areas, especially the north of the country versus the south. In particular, the sharp contrasts in the work ethics of the north and the south have been a vexing subject for many Italian politicians. Two Italian researchers, Andrea Ichino and Giovanni Maggi, studied a database containing information on the behaviors of thousands of employees at one of Italy's largest banks.[7] The collected data on each employee included detailed information on the number of times the employee came to work late or failed to show up entirely, promotions to higher ranks in the bank hierarchy, and transfers from one branch to another. Using this information it was possible to identify bank employees who had moved from bank branches in the north to branches in the south and vice versa.

Ichino and Maggi discovered that bank employees who moved, for example, from Milan in the north to Naples in the south, exhibited extreme changes in their work behavior. Once in Naples, they were frequently late to the office and missed significantly more days of work. Since only sick days are officially considered acceptable reasons for missing days of work at the bank, one might surmise that the move from one city to another led to deteriorations in the health of transferred employees, but the researchers also discovered that employees transferred from Naples to Milan also exhibited different patterns of behavior, which in this case was expressed in lower rates of tardiness at work and fewer missed work days.

Further analysis of the database proved in a convincing manner that the only reasonable explanation for these changes in behavior patterns was the phenomenon of peer effects. Employees transferred from Milan to Naples very quickly learned that their colleagues in Naples had a weaker work ethic than the one they had been used to in Milan. This reduced the internal incentives they had to maintain the high standards of work ethic that they had been used to in Milan. In contrast, employees transferred from Naples to Milan learned (although apparently not as quickly as those moving in the opposite direction) that they were now in a very different work environment, in which their colleagues invested more time and energy at their jobs. This, of course, was an uncomfortable position for them, but it still created an incentive for them to adopt the work ethic of those around them.

In follow-up work, Ichino, this time in collaboration with Armin Falk of the University of Bonn, studied peer effects in a laboratory setting.[8] In an early version of Ichino and Falk's experiment, subjects (students) were employed in preparing invitation letters to a (fictitious) conference that was to take place in Zurich. Subjects were hired to work for fixed time periods at a set wage ($20 per hour). Their task was to fold invitation letters, insert them into envelopes, seal the envelopes, and stamp them.

The subjects of the experiment were divided into two groups, with each group in a separate room. The experimenter visited the first room several times over the course of the experiment, each time placing on a prominent table in the center of the room a thick parcel of envelopes ready to be stuffed with more invitations. In contrast, he entered the second room a relatively small number of times, bringing into the room only a small stack of envelopes on each visit.

The results of the experiment indicated that in the first room, where subjects frequently saw large batches of envelopes placed in front of them, the students made more of an effort to work hard than in the other room. As in the other experiment, Ichino and Falk explained the difference in the observed behaviors in the two rooms as due to peer effects. The subjects in the first group were given the impression that their peers were working hard, stuffing the envelopes at a rapid pace. A potential sense of shame if their accomplishments were to fall too far from the standards of their peers spurred these students to work particularly hard. Interestingly, the peer effect appeared here despite the fact that the subjects did not know one another.

In the other group, the effect went the other way. With only small numbers of new envelopes being introduced into the room, the subjects received the impression that they were managing to stuff more envelopes than the others. In other words, subjects who worked too hard started feeling like they were "suckers." This is an unpleasant feeling, and they therefore slowed down their pace of work to avoid experiencing it. This experiment showed convincingly that it is not only monetary incentives that have effects on worker behavior—social incentives also play a significant role.

The third and last example presented in this chapter is, like the bank employees example above, based on data collected in real-life situations, not lab experiments. Two Berkeley researchers, A. Mas and E. Moretti, studied peer effects among checkout counter workers at a large American supermarket.[9] Many work actions conducted by checkout counter workers are routinely recorded in computer databases (such as the start and end times of scanning each customer's checkout items, and the number and types of items that a customer has brought to checkout). Using these figures, Mas and Moretti were able to estimate the efficiency of each worker in terms of number of items scanned during a given period of time. They discovered that when one worker ended a shift and was replaced by a different worker, the shift handover affected all the workers who were close enough to observe it. If the worker beginning a new shift was more efficient than the worker that he or she had replaced, nearby workers picked up the pace of their work. But if the new worker was slower, the others also slowed their pace accordingly.

These three research studies show that teamwork effort is significantly affected by the work environment. Each individual's motivation to invest in harder work is increased if those around him are working hard. In the mathematical model presented above, this occurred because we assumed that the success of a project required that both employees put in extra effort, but this is not necessary in more general settings: psychologically, workers want to avoid feeling exploited on the one hand, while on the other hand they did not want to be perceived as exploiting their coworkers.

There are many other studies indicating that psychological and social elements are critically important to getting team incentives to work properly between peers. When interactions between employees at different levels of corporate hierarchies are studied, particularly those between supervisors and teams working under them, it turns out that the psychological and social elements can have even more decisive effects.

Unfortunately, the business world so far seems to have ignored a good deal of what has been learned in research studies focused on teamwork in particular and on incentives in organizations in general. Human resources directors and organizational consultants continue to stress monetary incentives (mainly bonuses) based solely on individual accomplishments, without considering team-based incentives. They do so not necessarily because they believe that team-based incentives may reduce worker efficiency, but for the simple reason that using standard and familiar methods based on purely individual incentives protects them from criticism if problems arise.

Since individual incentives require detailed monitoring of each individual employee's effort, which is usually not feasible, human resources departments tend to use evaluation metrics that are often almost entirely irrelevant. This prompts employees to invest in optimizing their human resources evaluation scores instead of the job they were hired to do. Determining bonuses on the basis of such weak metrics distorts the system of incentives and furthermore engenders feelings of frustration, unfairness, and envy among employees, often leading to conflicts within teams instead of harmonious teamwork.

One of the most common evaluation metrics used by human resources departments is based on horizontal peer reviews, in which employees are asked to evaluate their colleagues' work efforts and at times even rank their colleagues. In competitive work environments, in which the value of team cooperation wilts in the intense heat of each employee's burning ambition to stand out in personal achievements, it should be self-evident that the trustworthiness of such reviews is extremely limited. Under this system employees have incentives to belittle the accomplishments of coworkers who threaten their positions while praising others who can later reward them. The reviews are subjective in any case, so this can be done without the employee being accused of lying.

A much more efficient incentive system is one that rewards successful teamwork while adding an additional but small bonus to the most industrious worker in the team. This harnesses together both a sense of group responsibility and a drive for personal achievement. In the description of the successful Continental Airlines incentive system that opened this chapter, it was not the offer of a gross benefit of $65 to each employee that was the key element to the project's success, but rather it was the responsibility that the employees felt toward each other, with each employee wishing to avoid blame for denying the bonus from his or her coworkers.

Several weeks ago I was speaking with my mother on the telephone when I had to apologize for cutting the conversation short because I had promised to take several friends of my son to a basketball game in which they were playing. My son was ill that particular day and therefore could not join us. "Make sure everyone has their seatbelts on," insisted my mother. "You are responsible for children that are not yours." Later in the car I pondered what she meant. I had never heard her say anything similar when I was alone with my son in the car. She could not possibly be less concerned about her grandson than about his friends. Perhaps she trusted that my paternal protection instincts would naturally kick in when I was alone with my son and wanted to make sure I would follow those same instincts when his friends were in the car without him?

Eventually I understood that my mother's admonishment stemmed from a moral imperative that requires a person to act especially carefully when he is responsible for children that are not his. A less extreme version of this same imperative is doubtless familiar to most of us. Imagine that it is your turn to take your child and your neighbor's child in your car to their kindergarten when you discover that there is only one child seat in the car. Which child will you place in the child seat? Less drastically, consider the fact that we often avoid asking a friend to lend us an item out of fear that we will somehow damage it before we can return it. And if it does get damaged for some reason, we then feel ten times worse than if the same damage were to occur to an item we own.

Exactly the same moral mechanism comes into play when collective rewards are at stake, which is why they are so powerful. An employee working as part of a team is less concerned about losing her own bonus as a result of laziness on her part than about how it will look if her laziness costs her friends the bonus that they are entitled to.

PART V
On Rationality, Emotions, and Genes

22

IRRATIONAL EMOTIONS

In one of the initial chapters of this book I described how anger can serve as a mechanism for creating credible commitment, enabling us to improve our strategic positions in interactions with others. Aristotle, who was quite aware of the important role that anger plays in our lives, wrote in his book *Politics* that "anybody can become angry—that is easy, but to be angry with the right person and to the right degree and at the right time and for the right purpose, and in the right way—that is not within everybody's power and is not easy." However, although anger is intended to benefit us from an evolutionary perspective, often it also harms us—not only because of the mental suffering that anger can cause, but also because of the implications it can have on our relationships with those toward whom we express our anger. We are often limited in our ability to control our anger in situations in which it does not serve us or even harms us.

Other emotional reactions with evolutionary advantages that have developed in the human race can similarly create social barriers or trip us up when we need to make correct decisions. In some cases, the evolutionary advantage of certain emotions can be overwhelmed by their disadvantages in the modern world. Several thousands of years of further evolutionary development may be required until they disappear entirely.

Blushing provides a very interesting example. Blushing is prompted by a sense of embarrassment, which is definitely a social emotion. When we feel shamed or embarrassed, the last thing we want to do is draw attention to ourselves. If we could, we would instead prefer to become invisible at those moments. And yet it is precisely in those situations that nature has chosen to highlight our presence by making our faces look as if a bright red light is shining on them.

Charles Darwin devoted an entire chapter in his book *The Expression of the Emotions in Man and Animals* to the subject of blushing. He identified it as one of the unique characteristics of the human species. But researchers specializing in the evolution of human psychology are still divided over the evolutionary source of blushing. Many regard it as a reaction of the sympathetic nervous system in preparation for what is called a "fight or flight" reaction. Pressure-filled and threatening situations stimulate increased blood flow to the head, because body tissue engorged with blood becomes more sensitive than usual, acting as a radar system to warn of impending danger. A side-effect of this is the red facial coloration of a blush.

This explanation is supported by an interesting experiment performed in 2003 in Australia.[1] Subjects in the experiment were asked to sing or read aloud a passage while they were viewed in profile—meaning that only one half of the face of each subject was visible to others. The experimenters found that blood flow to the half of the face exposed to the gaze of others was higher than blood flow to the other half. Blushing, in other words, was localized and appeared in the part of the face that was the most exposed to "danger."

An alternative explanation of the evolutionary advantage of blushing focuses on the reliable signal that the blusher sends out to the social environment, assuring others that the fact that an unacceptable action or a deviation from social norms has occurred is being duly acknowledged by the blusher. This message, which is reliable precisely because blushing is not given to our control and cannot be consciously faked, served the interests of blushers in the past by making social punishment redundant. Recently conducted empirical studies show that individuals who violate social norms and blush as a result are regarded less negatively by others than those who do not react by blushing. Blushing, however, also occurs in other situations, such as when we are the subjects of overflowing praise. In those cases blushing is socially less advantageous than not blushing.

Regret, which has very clear evolutionary advantages, is also an emotional reaction that can have negative effects, sometimes leading us to make suboptimal decisions. If we never felt regret for any of our actions, we would doubtless be quite miserable, doomed to repeat the same mistakes over and over again. Bonnie Ware, a palliative care worker with many years of experience treating terminally ill patients in hospices, wrote a book about the five most common and sharply felt regrets of the dying as they expressed them to her in their final weeks of life.[2] Men typically regretted overconcentration on their jobs throughout their lives and the old friendships that they had lost over the years. Women regretted not giving themselves permission to be happy often enough and also spoke of investing too much in pleasing others. Both sexes regretted holding back on expressing their feelings to others.

It would seem that these regrets are almost by definition irrational emotions—because they were expressed by people who knew that they were so close to their deaths that they could not possibly have sufficient time left to change their behaviors significantly. But these sorts of "grand" regrets actually are in most cases (though not on the cusp of death) quite rational emotions. They are usually most keenly felt during life-changing crises that provoke us to conduct major reviews of our lives and the directions they are taking. They can often bring about significant changes in our habits that last long after the crises that sparked them have abated.

The irrational regrets are most often the smaller ones that cause us to take biased decisions before we have learned all the facts and checked our actions rationally. A large number of experiments conducted by economics and finance researchers have shown that we tend to act with the goal of minimizing future regret. One example of behavior that is guided by a desire to reduce regret is the conformism mentioned in the chapter on herd behavior. We tend to adapt our choices to resemble those chosen by most of our acquaintances. If, for example, the majority of our friends have sold all their stocks out of fear that the stock market is about to collapse, we tend to do the same, even if we are getting strong objective indications that the market is going to rise in the near future. We do so because our regret will be less keenly felt if our mistake is shared by all our friends than if we are left all alone in making the wrong decision. For similar reasons, we are often more fearful of making decisions in subjects that we are supposed to know well and less sensitive to risks in areas in which we have no understanding. We strive to avoid, at almost all costs, the first type of regret (associated with a wrong decision on an issue we are supposed to be familiar with) even if undertaking the risk is worthwhile.

Fear of feeling regret sometimes makes us stubbornly stick to an erroneous decision to avoid admitting we made a mistake. For example, we sometimes find it difficult to sell an asset or an investment instrument in which we have lost money because doing so is tantamount to an admission that we made a mistake investing in it in the first place. As long as we continue to hold on to the asset there remains a chance that we will not regret buying it, because its value might still rise. This leads many to hold losing assets long after it has become unreasonable to expect that their worth will ever return to what it was when they bought them.

Georgio Coricelli, a University of Southern California researcher, along with several coathors, conducted a thorough study about a decade ago on brain activity related to feelings of regret.[3] In contrast to most other emotions, the brain activity that is registered during feelings of regret is spread out among several parts of the brain, from those related to cognition and analytic thought such as the orbitofrontal cortex and the inner sections of the cortex to parts of the limbic system such as the hypocampus, which regulates emotions and memories. This wide use of so many parts of the brain may be due to the nontrivial learning aspect of regret, beginning with analytic activity intended to measure to what extent regret for our actions is "justified."

Coricelli and his coauthors discovered that the brain activity that occurs when we try to minimize future regret in our decisions is very similar to brain activity when we experience regret itself. It would appear that when striving to minimize regret, we focus on the negative outcomes that our decision might cause, and it is during this process that we experience future regret.

There are several other types of irrational economic behaviors that are linked to specific brain activities. Many of them are related to dopamine, the "incentive hormone," and to the absorption rate of that hormone in the brain. Earlier chapters noted that this hormone is involved in the feelings of satisfaction and pleasure we get from success and that it can influence our attitudes toward risk. It incentivizes us toward achievement, which has clear evolutionary advantages. Dopamine is also responsible for several other brain functions; lack of dopamine is related to Parkinson's disease. Our need for the feelings of satisfaction and pleasure that dopamine gives us can cause us to act in ways that are inimical to our material interests and in some cases can even be the source of mental disturbances. Kleptomania, oniomania (shopping addiction), and ludomania (gambling addiction) are well-known psychiatric conditions related to perverse economic behavior. In certain cases the psychiatric treatment given to people suffering from these conditions involves balancing dopamine levels in the brain.

But even under normative behavior dopamine can cause us to make erroneous decisions based on emotions. One of the most prominent of such phenomena can be seen in the behavior of auction participants. In recent years tenders and auctions have proliferated extensively in cyberspace. The amount of money traded in online auctions has correspondingly grown to astounding proportions. In 2000 the *New York Times* assessed the mobile telephone spectrum auction conducted that year in Britain to be the largest auction ever held in history. That single auction raised more than $34 billion in revenues from the sale of mobile telephone frequencies.

One of the most intensely studied phenomena associated with auctions is known as "the winner's curse": in many cases the winner of an auction actually pays more for the item he or she has won than its true worth. This is a phenomenon observed not only in auctions of low-priced items whose participants are amateur bidders; large corporations bidding in major tenders are also prey to the winner's curse. In the early 1970s many US oil companies collapsed shortly after they won auctions granting them drilling rights in several places in the United States. These corporations had large staffs of geologists and economists assessing the values of the drilling rights for which they were bidding, but it turned out that they had bid prices that were much higher than the true values of the drilling rights on offer, which eventually bankrupted them.

There are two main causes of the winner's curse, one cognitive and one emotional. Participants in an auction try to assess the value of the item on auction as best as they can. They then submit initial bids that are slightly lower than that assessment. The more competitive the auction environment, the closer bids will be to the assessed value, because the more bidders there are in an auction, the higher the chances that someone else will outbid you.

If there are a very large number of bidders and they have conducted independent value assessments, it is reasonable to suppose that the average assessment will be quite close to the true value of the auctioned item. If that's the case, then the winner of the auction, who has bid the most money, has made an offer higher than the average bid—meaning that it is probably higher than the true value of the auctioned item. This is the cognitive explanation for the winner's curse. In other words, the participants fail to take into account the fact that if they submit the winning bid, then they are valuing the auctioned item higher than everyone else, which in turn means that they are likely to be overvaluing it.

One way to avoid cognitively falling prey to the winner's curse is to write down the price you are willing to bid on a piece of paper that you then stow away in a drawer for twenty-four hours. After twenty-four hours have passed, take it out again and imagine that one of the auction officials who has already seen all the other bids informs you that you have submitted the highest bid. You should now recalibrate your bid based on this information. In most cases, this will lead you to lower your bid, protecting you from the winner's curse.

But there is also an emotional cause to the winner's curse in many cases. Participants in auctions often find themselves driven to submit high bids by "auction fever"—an uncontrollable desire to win the auction at any price. A few years ago two students of mine asked me to suggest a research project. My advice was to find Web sites in which the same items were offered for sale in two different ways—by auction versus direct sale at a fixed price—and to compare the prices at which the items were eventually sold. I hypothesized that the auction prices would in many cases be higher than the direct sale prices for the same items, and my hypothesis turned out to be true. The auction participants could have obtained what they had bought by auction at lower prices if they instead had gone to direct sales at the same site, but the competitiveness of the auction environment and the auction fever that accompanies it pushed them to pay much more.

I personally witnessed a very expensive instance of auction fever several years ago when I was involved in the planning stages of the auctioning of the State of Israel's national gas storage company. Four large private oil and gas corporations competed vigorously and competitively in that auction, with the winning bid totaling $220 million—almost twice the amount that we expected to receive under even the most optimistic estimate supplied to us by assessors prior to the auction.

When I was hired as a consultant to one of the companies participating in the bidding in Israel's mobile telephone spectrum auction in 2011, I strove to help my client avoid falling into the trap of the winner's curse. A few hours before the auction began, I advised the company owner to step away from the overexcited tumult raging at corporate headquarters, take a deep breath, and find a quiet and relaxed place in which he could calmly assess the highest possible price he would truly reasonably be willing to pay for the asset he was seeking. I further told him that after arriving at his decision, he should write that amount of money on a piece of paper and insert it into a sealed envelope. The envelope would then in turn be entrusted to one of his closest friends, a senior banker who was present at corporate headquarters that day. Doing so would constitute a commitment to refuse to be tempted into making a bid beyond that predetermined maximal amount.

The company owner, an experienced, brilliant, and talented businessman, was taken aback by my request. He told me not to ask him to do anything he couldn't commit himself to doing. He agreed to my suggestions only after repeated entreaties from his colleagues that he would take my advice seriously, along with a threat by the banker to leave the premises if my instructions were not carried out.

The computerized auction began at 11 a.m. and continued straight through to 8 p.m.—nine nail-biting hours of nerves. At 7 p.m., with the latest bid price at $135 million, the owner slowly rose from the chair in which he was sitting, stretched out to his full height, and took a long gulp from the cup of coffee in his hand. "My dear friends," he admitted somewhat shamefacedly, "the truth is that I bid well beyond the highest amount that I wrote down in the envelope."

He continued to increase the company's bids over the remaining hour of the auction, down to the wire.

When the dust settled, it turned out that he had won the auction by submitting a bid of almost $200 million—just about twice as much as the absolute maximal amount he had declared the asset was worth on the piece of paper inserted into the envelope prior to the start of the auction. Forty-five days later, his winning bid was retroactively nullified in accordance with the rules of the auction because he was unable to obtain sufficient bank guarantees to support such an exorbitant commitment of money.

Dopamine plays a role in the emotional aspect of the winner's curse. Several studies have been conducted using fMRI brain imaging to track the brain activities of auction participants.[4] These revealed complicated patterns of brain activity involving several different parts of the brain, but one phenomenon in particular stood out as subjects exhibited auction fever. At every stage in which subjects were informed that they had failed to win an auction, subactivity was registered in the striatum area. This area is part of the limbic system and a site at which dopamine is secreted in the body. Prominent auction losses prompted more prominent subactivity in the striatum, which was followed by more aggressive bidding on the part of subjects in the next stage.

There is no method for completely eliminating irrational emotions, but they can be dulled, and their negative effects can be reduced by conscious awareness of their existence and effects. I have stressed throughout this book that the cognitive and emotional systems within us are not entirely separate from each other and that they often work together. What determines whether a particular emotion is working for us or against us is very much tied to the circumstances that generate that emotion. Identifying the effects of emotional reactions is in many cases a task that our cognitive faculties need to undertake. Just as our cognitive system can enhance emotions that are helpful to us, it can rein in those emotions that are inimical to our best interests. The quote by Aristotle that appears in the initial paragraph of this chapter is right on the mark on this issue. Controlling emotions, especially anger, is not easy. It requires analysis, memory, intuition, and skill, but it does pay off.

23

NATURE OR NURTURE

What Is the Source of Rational Emotions?

SOME YEARS AGO I RAN INTO AN OLD SCHOOLMATE OF MINE, OFER Lipschitz. "We searched for you months ago to invite you to a school reunion held in my home," said Ofer apologetically. "Someone mentioned that you were living abroad, so we didn't really make much of an effort to locate you." Ofer and I, along with all the attendees at the reunion, were in school together for eight consecutive years, from first grade at age six and onwards. When I told Ofer how disappointed I was at having missed the reunion, he tried to comfort me: "We filmed the entire event. Till the next time we conduct a reunion, you can at least watch the video." The video Ofer gave me included no less than three hours of footage that contained many shots of every single reunion participant.

Two things especially moved me while I watched the video. The first was the fact that nearly everyone (including me) managed to identify each reunion participant as soon as he or she walked through the door, even before they introduced themselves. We were able to do this despite the fact that the last time we had seen each other was thirty-five years earlier, when we were children. This amazing capacity we have is related to the deep traces that facial features leave in our memories. If I were shown a class photograph of other children from the same time period next to pictures of those same children today as adults, I doubt that I would be able to make even one successful match. The way that facial features are stored in our brains apparently differs from the way other information is stored. We often meet people who seem familiar to us, or whom we are certain we have previously met, but we find ourselves unable to recall any details about them—neither their names nor where and when we met them.

The other thing that impressed me while viewing the reunion video was that the facial features of my friends were not the only aspects of them that were entirely recognizable to me, based on my childhood memories. The present occupations of many of them seemed to be entirely predictable. Ofer and Myron, who were entertaining us with their guitar playing at our get-togethers as far back as the fifth grade, had developed musical careers, deriving their incomes from musical performances and music instruction. Tali, who early on had exhibited much more interest in boys than any other girl in the class and had always been the main source of information on who was romantically attached to whom, is now a sexologist and marriage counselor. Yossi, who had initiated and organized most of our social activities as schoolchildren had grown up to be an entrepreneur, founding and directing start-ups.

Personality traits had also been preserved to a surprising extent from childhood to adulthood. Those who had been introverted as children stood apart and alone at the reunion, looking a bit out of place in the socially intense event taking place around then. Those who had laughed often as children laughed just as often as adults, the loud kids had become loud adults, and the handful who had been prone to antisocial violence as children did not show up at all.

Anyone who experiences such a reunion cannot fail to come away with the strong insight that major elements of personalities are determined in the first ten years of our lives. In fact, in recent years an increasing number of scientific studies are revealing that our personalities are fashioned even earlier than that—not in the first ten years after birth but in the nine months preceding it. The full mapping of the human genome is a huge leap forward for un-covering new insights in what determines personalities. One after another, new discoveries are exposing the tight connection between specific personality traits and genetic profiles.

Richard Ebstein, who specializes in genetic psychiatry at the Singa-pore National University, has conducted a number of very interesting studies on this subject matter, along with several coauthors. In one of these they focused on the genetic basis of generosity.[1] As noted in an earlier chapter of this book, oxytocin is responsible for the mutual empathy felt by mothers and their newborn infants. Another hormone that plays a role in creating the mother-child bond is vasopressin, which is important for several emotional and physiological human functions. The main gene responsible for secreting vasopressin is called AVPR1a, which exists in different lengths. Shorter versions of this gene, which tend to create smaller amounts of vasopressin, are more common among individuals suffering from autism spectrum disorder.

Ebstein and his colleagues studied hundreds of healthy subjects and categorized them according to the length of this gene that each one carried. They then had these subjects play the giving game, as described in Chapter 9 (in the giving game, each player receives a sum of money and can donate as much as he wishes from that sum to the other player). Subjects carrying shorter versions of the gene gave much less to others in this game than those with longer versions. This was a clear identification of a (statistically observable) personality trait with the variations of a single gene, in this case AVPR1a.

Other research studies based on comparisons of the behaviors of identical twins have uncovered the genetic sources of several other personality traits. If a certain personality trait is exhibited with high correlation between identical twins (who share the same genetic pro-file) and low correlation between fraternal twins, that serves as an indication that the genetic component of that trait is more significant than the social component.

Richard Ebstein and his colleagues also conducted a wide-ranging survey of research studies of personality traits and used that to estimate the genetic components of each.[2] Figure 2 presents a summary of their results, with two different types of genetic components presented separately: purely genetic components that are independent of social influences are labeled DZ, and more general genetic components—which take into account possible social influences but only those that influence individuals with particular genetic profiles—are labeled MZ.

Figure 2 shows that there is a significant genetic component to a wide range of behavioral traits. In some cases this component is dominant. Most of the research studies on which the table is based are quite recent. The data in those studies, along with additional indications that genetics play a major role in determining personality traits, are reviving the age-old debate on nature versus nurture, which in the past has too often conflated moral considerations with scientific claims.

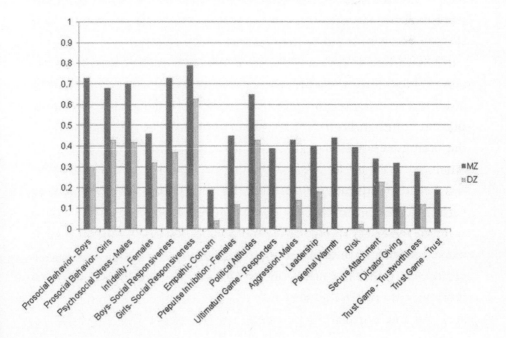

There is some cause for concern given the potential implications of the research studies mentioned in this chapter. It is now possible to conduct a complete mapping of the genetic code of an individual at a lower cost than ever before. A sample of saliva and $200 are sufficient to identify our genetic tendencies. The more we understand about the genetic component in determining personality traits, the more the private market will be incentivized to make use of the information stored in our DNA. Job applicants may one day be asked to provide saliva samples along with their resumes. Home renters could similarly ask potential tenants for their saliva samples, and insurance companies could use genetic profiles to calculate insurance premiums. This may eventually extend to all our economic and contractual interactions, creating a self-reinforcing cycle: those with "attractive" DNA profiles getting the best jobs, and those with "ugly" profiles being forced to work in more unappealing occupations, or suffering unemployment. Within the span of a few years this sort of discrimination could in itself strengthen the impression that a person's DNA profile categorically determines his or her chances of success in life. Those stigmatized with "ugly" DNA profiles would have no incentive to seek higher education, learn a profession, or even work hard. Social mobility would slow to a trickle, as a new genetic aristocracy emerges.

Despite these clear dangers, we should not use our fears as an excuse to block future scientific research. Ignorance should never serve as a form of inoculation against potential social dangers. It is already becoming nearly impossible to obtain grants to study the relationship between genetics and cognitive ability or IQ from the major national scientific funds in the United States, the National Science Foundation and the National Institutes of Health. There is no formal decree forbidding such research, but political correctness seeps into deliberations on research funding early on in the process, dooming most such grant requests. Concerns about the potential use of such research efforts to justify discrimination are understandable, but self-imposed ignorance is no solution either.

EPILOGUE

SOME OF MY READERS MAY HAVE BEEN UNCOMFORTABLE WITH THE prosaic approach expressed toward emotions in this book. If emotions, as presented here, are as rational and analytical as cognitive processes, what room is there left for the soul? Furthermore, if the most nonphysical aspect of the human experience, our "emotions," can be predicted by DNA and the concentrations of certain hormones in the body, then could we not use similar descriptions to capture the entirety of what we call "life" and do away once and for all with the concept of spirit?

I don't think so.

Scientific understanding enables us to capture only a partial and somewhat faint picture of the totality of human emotion and cognition. The full picture is far from being entirely clear and may never be completely clarified. Spirit and soul represent that which is hidden away from what we can explain scientifically.

The weightiest philosophical questions on the essence of the soul and of life remain unresolved. Are we, who are composed of biological cells containing DNA macromolecules that are themselves composed of carbon and hydrogen, only the sum of our parts, or is there, over and above the material composition of our bodies, an additional mysterious ingredient that distinguishes the living from the inanimate world? Is there an ingredient without which any synthetic creations that we can conjure up in a laboratory will never truly be alive?

In every scientific explanation, from the most fundamental physics through to biology and on to economics, the question of whether or not the explanation represents absolute truth is almost irrelevant. A scientific explanation is admissible if it succeeds in consistently and efficiently explaining empirical phenomena that we discover around us. Quantum theory and the theory of relativity are sufficiently simple for physicists to be convinced that they explain a vast number of physical phenomena that would otherwise be inexplicable (even though you and I may struggle to understand those theories ourselves). Yet they may be very far from the "truth," simply because we are technologically and cognitively limited in what we can observe around us. The physical model of antiquity, which posited that the world is a flat plate beyond which lies an unfathomable abyss, is an excellent model given the empirical observations that were available when it was believed. It only ceased to be a persuasive model when seafarers returning from distant journeys told of observations that were at odds with its explanation.

The same applies to scientific explanations of human emotional and cognitive behavior. Game theory, the brain sciences, evolutionary explanations, and psychology are but a story (or several stories) intended to help us make sense of our observations of individual and group behaviors in various situations. In contrast with physical theories, behavioral theories, especially those based on game theory, have in recent years been developing at a rapid pace. This is because empirical outcomes in the behavioral sciences are much more accessible than their counterparts in physics. Neither giant telescopes nor particle accelerators are necessary to attain empirical behavioral outcomes. In the behavioral sciences, standing in a queue at the supermarket or reading a newspaper item can serve as an empirical workshop, stimulating insights in the minds of researchers that can be tested in brief and relatively inexpensive lab experiments. New insights that are empirically confirmed accumulate until they become the foundations of new stories (also known as theories) that slightly improve the resolution of the general picture that we have of what is termed "human behavior."

The ease with which new empirical outcomes can be generated is a major advantage that accrues to behavioral research, but it is also potentially a danger for the field. While laboratories in the physical sciences can provide us with objective measurements of physical constants (in some cases up to ten decimal places), behavioral laboratories generate outcomes that may at times be given to disparate interpretations. These outcomes are sensitive not only to how the experiments were planned and executed, but also to the way the resultant data are analyzed. Researchers whose intellectual integrity may not match the most rigorous of standards and who are eager to obtain particular experimental results have been known to "torture" data until it "confesses" what the experimenter wants to hear. These sorts of actions expose behavioral laboratory results to potentially damaging manipulations. As competition between leading researchers becomes increasingly aggressive, this danger becomes ever more present.

In 2011, Tilburg University in the Netherlands terminated the employment of Diederik Stapel, a prominent professor of social psychology who was the dean of that university's School of Social and Behavioral Sciences. That action followed revelations that the up-and-coming academic star had for years blatantly fabricated the data on which he had based his research work. Dozens of papers published by Stapel in leading peer-reviewed journals had to be retracted.

One of Stapel's research efforts even attracted interest in daily newspapers throughout Europe. In that paper, titled "Meat Gets the Worst Out of You," Stapel claimed that eating meat, or even thinking about it, turns people into selfish antisocial individuals. He based his conclusions on empirical data that he had ostensibly collected in laboratory studies, but it turned out that all the data had been entirely generated by his fertile imagination.

Investigation of Stapel's fraudulent actions and his subsequent admission that he had conducted behavior exceptionally unworthy of a scientist cost him his career. He was removed from his position as a professor at Tilburg University and even expelled from membership in the academic societies to which he belonged. The head of an international psychological association that expelled Stapel composed a letter sent to all the members of the association warning that overly aggressive academic competition can cause people to lose their minds and commit fraud.

The entire episode itself raises interesting questions about human nature. Academic competition promises its competitors neither money nor any other material benefit as its reward. It is a competition for recognition, respect, and praise. It turns out that people can derive a great deal of pleasure from recognition and praise even when they know that they are entirely unworthy of it. Stapel's fraudulent activity is probably not the only case of a researcher manipulating data, but it is a rare example of such blatant behavior. Academic checks and balances usually operate quite efficiently to uncover such cases, but a dose of critical thinking and skepticism is a healthy thing to maintain with respect to every behavioral research effort, including those mentioned in this book.

✳ ✳ ✳

THE BORDER DELINEATING THE FRONTIER BETWEEN OUR INTERNAL emotional and rational systems is very thin and convoluted. In most of the occasions in which we are called upon to make decisions, whether those are monumental life-changing decisions or the most minor and mundane ones imaginable, that thin line is liable to become so blurred that it may disappear entirely. The two systems become intertwined around each other so tightly that they become inseparable. In many cases our emotions are there to enable us to arrive at rapid and nearly automatic decisions, but in other cases, especially when weighty issues are at stake, our emotions challenge our rational thought processes.

For decisions on subjects such as which job to seek or whether to continue a romantic relationship, it is not our rational mechanisms that come to the fore but our emotional ones. We almost always reach a point at which all the facts are known to us, we have gone over all the implications of each alternative several times, we know that there is no new item of information or new insight that is likely to arrive and help us make up our minds, and yet we still find ourselves unable to take that last step and decide. What holds us back is emotional, not cognitive. Rational considerations (and even material interests) translate into emotional reactions—fear versus hope, sometimes compassion versus anger, pull us in different directions like a swinging pendulum, until finally the decision comes down to which emotion is most deeply felt. That is how our decision-making "software" really works, and in most cases that is a very good thing.

Try for a moment to imagine what would happen if a "separation of authorities" were to apply to our decision making in the same way it is applied to the branches of our governments, with some decisions handled solely by our rational mechanisms and others controlled only by our emotions. To take a concrete example, consider a situation with which many of us are familiar. You arrive at your place of work one morning, switch on the computer, and find an email message from another company suggesting that you apply for a position there. It is up to you to decide whether to send a positive response to this offer or instead to decline politely and continue in your present job.

If this decision were to be relegated solely to the rational department of your mind, you would react as one might expect Mr. Spock from the planet Vulcan would react. You would begin by compiling an exact list of all the characteristics of your present job (salary, personal interest, opportunities for promotion) followed by a parallel list of the characteristics of the newly offered job. Since you have only partial information available regarding the job on offer, you will assign probabilities to each of the characteristics in the list you have compiled for that position. You will also be able to predict quite accurately the chain of events that are likely to occur once you decide to apply for the job, along with your chances of landing that job at the end of the process.

The next step would be to assign to each characteristic a value representing the extent of satisfaction or disappointment you expect to receive from it. If you are lucky and hitherto have not made any mistakes, you are almost certainly about to fail at this stage. There is almost no way to gauge values of satisfaction and disappointment without the assistance of your emotional mechanism. You will have all the facts at your disposal, but you will not be able to make a wise choice, after all that.

What would happen if instead you were to empower only your emotional mechanism to make the decision for you? You might then arrive at a decision quickly, but it would be dictated by recent events whose relationships to your long-term interests are tenuous at best. If, for example, on the previous day your boss had spoken to you in a way that irritated you, you might reply positively to the email inviting you to apply for another job and even include your boss in the list of recipients. On the other hand, if the person who sent you the email misspelled your name, or the proposed position looks at first glance as slightly less attractive than your current job, you would likely immediately reject it along with a sarcastic remark that would obliterate any chance that you would ever again be approached with a job offer. Only close cooperation between the emotional and rational mechanisms can enable you to arrive at a wise and satisfactory decision.

I hope that the examples and the many research studies surveyed in this book convince you that emotions are not a vestigial leftover of the evolutionary process from a long ago primitive past, but rather an effective and sophisticated tool for balancing and complementing our rational side. In the end, it is the feeling and thinking person who has the advantage, not the person who relies on thought alone.

NOTES

PREFACE

1. D. Ariely, *Predictably Irrational* (New York: HarperCollins, 2009).
2. D. Kahneman, *Thinking Fast and Slow* (New York: Farrar, Straus, and Giroux, 2011).

CHAPTER 1: WHAT IS THE POINT OF GETTING ANNOYED?

1. M. Tamir, "What Do People Want to Feel and Why? Pleasure and Utility in Emotion Regulation," *Current Directions in Psychological Science* 18 (2009): 101–105.

CHAPTER 3: EMOTIONAL IMPOSTORS, EMPATHY, AND UNCLE EZRA'S POKER FACE

1. M. Meshulam, E. Winter, G. Ben Shahar, and Y. Aharaon, "Rational Emotions in the Lab," *Social Neuroscience* 7, no. 1 (2012): 11–17.
2. G. McCarthy, A. Puce, J. C. Gore, and T. Allison, "Face-Specific Processing in the Human Fusiform Gyrus," *Journal of Cognitive Neuroscience* 9 (1997): 605–610.
3. A. Kalay, "Friends or Foes? Empirical Test of a Simple One-Period Division Game Having a Unique Nash Equilibrium," mimeo, 2003.

4. G. Rizzolatti and L. Craighero, "The Mirror-Neuron System," *Annual Review of Neuroscience* 27 (2004): 169–192.

CHAPTER 4: GAME THEORY, EMOTIONS, AND THE GOLDEN RULE OF ETHICS

1. B. Aumann and M. Maschler, *Repeated Games with Incomplete Information* (Cambridge, MA: MIT Press, 1995).
2. E. Winter, I. García-Jurado, and L. Méndez Naya, "Mental Equilibrium and Rational Emotions," Center for the Study of Rationality, Hebrew University, 2009.

CHAPTER 6: ON DECENCY, INSULT, AND REVENGE

1. W. Güth, R. Schmittberger, and B. Schwarze, "An Experimental Analysis of Ultimatum Bargaining," *Journal of Economic Behavior and Organization* 3, no. 4 (1982): 367–388.
2. Max Planck Institute, "Chimpanzees, Unlike Humans, Apply Economic Principles to Ultimatum Game," ScienceDaily, October 7, 2007.
3. A. E. Roth, V. Prasnikar, M. Okuno-Fujiwara, S. Zamir, "Bargaining and Market Behavior in Jerusalem, Ljubljana, Pittsburgh, and Tokyo: An Experimental Study," *American Economic Review* 81, no. 5 (1991): 1068–1095.
4. E. Winter and S. Zamir, "An Experiment with Ultimatum Bargaining in a Changing Environment," *Japanese Economic Review* 56 no. 3 (2005): 363–385.
5. A. G. Sanfey, J. K. Rilling, J. A. Aronson, L. E. Nystrom, J. D. Cohen, "The Neural Basis of Economic Decision-Making in the Ultimatum Game," *Science* 300, no. 5626 (2003): 1755–1758.

CHAPTER 7: ON STIGMAS AND GAMES OF TRUST

1. S. Knack and P. Keefer, "Does Social Capital Have an Economic Payoff? A Cross-Country Comparison," *Quarterly Journal of Economics* 112 (1997): 1251–1288.
2. GDP is a country's gross domestic product, the main index used to measure the economic development of nations.
3. J. Berg, J. Dickhaut, and K. McCabe, "Trust, Reciprocity, and Social History," *Games and Economic Behavior* 10 (1995): 122–142.
4. C. Fershtman and U. Gneezy, "Discrimination in a Segmented Society: An Experimental Approach," *Quarterly Journal of Economics* 116, no. 1 (2001): 351–376.

Chapter 8: Self-Fulfilling Mistrust

1. F. Bornhorst, A. Ichino, O. Kirchkamp, K. Schlag, and E. Winter, "Similarities and Differences when Building Trust: The Role of Culture," *Experimental Economics* 13, no. 3 (2010): 260–283.

Chapter 10: Collective Emotions and Uncle Walter's Trauma

1. G. Bornstein, E. Winter, and H. Goren, "An Experimental Study of Repeated Team Games," *European Journal of Political Economy* 12 (1996): 629–639.

2. G. Bornstein, E. Winter, and H. Goren, "Cooperation in Inter-group and Single-group Prisoner's Dilemma Games," in *Understanding Strategic Interaction—Essays in Honor of Reinhard Selten,* edited by W. Albers, E. van Damme, W. Güth, P. Hammerstein, and B. Moldovanu (Berlin and New York: Springer-Verlag, 1997), 418–429.

Chapter 11: The Handicap Principle, the Ten Commandments, and Other Mechanisms for Ensuring Collective Survival

1. A. Zahavi, "Mate Selection—A Selection for a Handicap," *Journal of Theoretical Biology* 53 (1975): 205–214.

2. R. Orzach, and Y. Tauman, "Strategic Dropouts," *Games and Economic Behavior* 50 (2005): 79–88.

3. J. Andreoni, A. Payne, J. D. Smith, and D. Karp, "Diversity and Donations: The Effect of Religious and Ethnic Diversity on Charitable Giving," NBER Working Paper 17618, November 2011.

Chapter 12: Knowing How to Give, Knowing How to Receive

1. U. Gneezy and A. Rustichini, "Pay Enough or Don't Pay at All," *Quarterly Journal of Economics* 115, no. 3 (2000): 791–810.

Chapter 13: The Spray That Will Give Us Love

1. E. Hart, S. Israel, and E. Winter, "Accuracy in the Perception of Social Deception Is Modified by Oxytocin," *Psychological Science* 25 (2013): 293–295.

Chapter 14: On Men, Women, and Evolution

1. D. Kahneman, A. B. Kruger, D. Schkade, N. Schwartz, and A. A. Stone, "Would You Be Happier If You Were Richer? A Focusing Illusion," *Science* 312, no. 5782 (2006): 1908–1910.

2. M. Francesconi, C. Ghiglino, and M. Perry, "On the Origin of the Family," discussion paper, University of Warwick, 2011.

3. M. Whitty and L. Quigley, "Emotional and Sexual Infidelity Offline and in Cyberspace," *Journal of Marital and Family Therapy* 34, no. 4 (2008): 461–468.

4. M. C. Neale, B. M. Neale, and P. F. Sullivan, (2002). "Nonpaternity in Linkage Studies of Extremely Discordant Sib Pairs," *American Journal of Human Genetics* 70, no. 2 (2002): 526–529.

5. U. Gneezy and A. Rustichini, "Gender and Competition at a Young Age," *American Economic Review* 94, no. 2 (2004): 377–381.

6. M. Niederle and L. Vesterlund, "Do Women Shy Away from Competition? Do Men Compete Too Much?," *Quarterly Journal of Economics* 122, no. 3 (2007): 1067–1101.

7. E. P. Lazear and S. Rosen, "Rank-Order Tournaments as Optimum Labor Contracts," *Journal of Political Economy* 89, no. 5 (October 1981): 841–864.

8. J. M. Coates, M. Gurnell, and A. Rustichini, "Second-to-Fourth Digit Ratio Predicts Success Among High-Frequency Financial Traders," *Proceedings of the National Academy of Science* 106, no. 2 (2009): 623–628.

9. D. Biello, "What Is the Best Age Difference for Husband and Wife?," *Scientific American,* December 5, 2007.

10. L. Brizendine, *The Female Brain* (New York: Morgan Road Books, 2006).

11. M. R. Mehl, S. Vazire, N. Ramirez-Esparza, R. B. Slatcher, and J. W. Pennebaker, "Are Women Really More Talkative Than Men?," *Science* 317 (2007): 82.

12. A. Christensen and C. L. Heavey, "Gender and Social Structure in the Demand/Withdraw Pattern of Marital Conflict," *Journal of Personality and Social Psychology* 59 (1990): 73–81.

13. L. M. Papp, C. D. Kouros, and E. M. Cummings, "Demand-Withdraw Patterns in Marital Conflict in the Home," *Personal Relationships* 16, no. 2 (2009): 285–300.

14. S. R. Holley, V. E. Sturm, and R. W. Levenson, "Exploring the Basis for Gender Differences in the Demand-Withdraw Pattern," *Journal of Homosexuality* 57, no. 5 (2010): 666–684.

15. U. S. Rehman and A. Holtzworth-Munroe, "A Cross-Cultural Analysis of the Demand-Withdraw Marital Interaction: Observing Couples from a Developing Country," *Journal of Consulting and Clinical Psychology* 74, no. 4 (2006): 755–766.

16. A. F. Bogaert, "Biological Versus Nonbiological Older Brothers and Men's Sexual Orientation," *Proceedings of the National Academy of Sciences* 103, no. 28 (2006): 10771–10774.

CHAPTER 15: MAKE ME A MATCH MADE IN HEAVEN

1. M. Perry, P. J. Reny, and A. J. Robson, "Why Sex? And Why Only in Pairs?," discussion paper, Center for the Study of Rationality, Hebrew University, 2009.
2. E. Illouz, *Consuming the Romantic Utopia: Love and the Cultural Contradictions of Capitalism* (Berkeley: University of California Press, 1997).
3. G. Becker, "A Theory of Marriage Part 1," *Journal of Political Economy* 81, no. 4 (1973): 813–846.
4. G. Becker, "A Theory of Marriage Part 2," *Journal of Political Economy* 82, no. 2 (1974): 11–26.
5. D. Gale and L. S. Shapley, "College Admissions and the Stability of Marriage," *American Mathematical Monthly* 69 (1962): 9–14.

CHAPTER 17: WHY ARE WE SO NEGATIVE?

1. J. von Neumann and O. Morgenstern, *Theory of Games and Economic Behavior* (Princeton, NJ: Princeton University Press, 1944).
2. R. C. Battalio, J. Kagel, and D. MacDonald, "Animals' Choices over Uncertain Outcomes: Some Initial Experimental Results," *American Economic Review* 75 (1985): 597–613.

CHAPTER 18: ON ARROGANCE AND HUMILITY

1. A. M. Spence, "Job Market Signaling," *Quarterly Journal of Economics* 87, no. 3 (1973): 355–374.
2. A. Tversky and D. Kahneman, "Extensional versus Intuitive Reasoning," *Psychological Review* 91 (1984): 293–315.

CHAPTER 19: OVERCONFIDENCE AND RISK

1. B. Barber and T. Odean, "Trading Is Hazardous to Your Wealth: The Common Stock Investment Performance of Individual Investors," *Journal of Finance* 55, no. 2 (April 2000): 773–806.
2. U. Gneezy, M. Niederle, and A. Rustichini, "Performance in Competitive Environments: Gender Differences," *Quarterly Journal of Economics* 188, no. 3 (August 2003): 1049–1074.

3. K. Dobson and R. L. Franche, "A Conceptual and Empirical Review of the Depressive Realism Hypothesis," *Canadian Journal of Behavioural Science* 21 (1989): 419–433.

4. M. Niederle and L. Vesterlund, "Do Women Shy Away from Competition? Do Men Compete Too Much?," *Quarterly Journal of Economics* 122, no. 3 (2007): 1067–1101.

CHAPTER 20: THE VOICE IS HERD

1. S. Bikhchandani, D. Hirshleifer, and I. Welch, "A Theory of Fads, Fashion, Custom, and Cultural Change as Informational Cascades," *Journal of Political Economy* 100, no. 5 (1992): 992–1026.

2. J. Bracht, F. Koessler, E. Winter, and A. Ziegelmeier, (2010) "Fragility of Information Cascades: An Experimental Study Using Elicited Beliefs," *Experimental Economics* 13, no. 2 (2010): 121–145.

3. B. Sacerdote, "Peer Effects with Random Assignment: Results for Dartmouth Roommates," *Quarterly Journal of Economics* 116, no. 2 (2001): 681–704.

4. S. Nieuwenhuis, B. U. Forstmann, and E. Wagenmakers, "Erroneous Analyses of Interactions in Neuroscience: A Problem of Significance," *Nature Neuroscience* 14 (2011): 1105–1107.

5. A. Tversky and D. Kahneman, "The Framing of Decisions and the Psychology of Choice," *Science* 211, no. 4481 (1981): 453–458.

CHAPTER 21: TEAM SPIRIT

1. M. Knez and D. Simester, "Firm-Wide Incentives and Mutual Monitoring at Continental Airlines," *Journal of Labor Economics* 19, no. 4 (October 2001): 743–772.

2. E. Winter, "Incentive Reversal," *American Economic Journal: Microeconomics* 1, no. 2 (2009) 133–147.

3. E. Klor, S. Kube, E. Winter, and R. Zultan, "Can Higher Bonuses Lead to Less Effort? Incentive Reversal in Teams," *Journal of Economic Behavior and Organization* 97 (2014): 72–83.

4. E. Winter, "Incentives and Discrimination," *American Economic Review* 94, no. 3 (2004): 764–773.

5. S. Goerg, S. Kube, and R. Zultan, "Treating Equals Unequally: Incentives in Teams, Workers' Motivation and Production Technology," *Journal of Labor Economics* 28 (2010): 747–772.

6. A. Cabrales, R. Miniaci, M. Piovesan, and G. Ponti, "Social Preferences and Strategic Uncertainty: An Experiment on Markets and Contracts," *American Economic Review* 100, no. 5 (December 2010): 2261–2278.

7. A. Ichino and G. Maggi, "Work Environment and Individual Background: Explaining Regional Shirking Differentials in a Large Italian Firm," *Quarterly Journal of Economics* 115 (2000): 1057–1090.

8. A. Falk and A. Ichino, "Clean Evidence on Peer Effects," *Journal of Labor Economics* 24, no. 1 (2006): 39–58.

9. A. Mas and E. Moretti, "Peers at Work," *American Economic Review* 99, no. 1 (2009): 112–145.

Chapter 22: Irrational Emotions

1. P. D. Drummond, L. Camacho, N. Formentin, T. D. Heffernan, F. Williams, and T. E. Zekas, "The Impact of Verbal Feedback about Blushing on Social Discomfort and Facial Blood Flow During Embarrassing Tasks," *Behavior Research and Therapy* 41, no. 4 (2003): 413–425.

2. B. Ware, *The Top Five Regrets of the Dying: A Life Transformed by the Dearly Departing* (Carlsbad, CA: Hay House, 2012).

3. N. Camille, G. Coricelli, J. Sallet, P. Pradat, J. R. Duhamel, and A. Sirigu, "The Involvement of the Orbitofrontal Cortex in the Experience of Regret," *Science* 304, no. 5674, (May 2004): 1167–1170.

4. M. R. Delgado, A. Schotter, E. Y. Ozbay, and E. A. Phelps, "Understanding Overbidding: Using the Neural Circuitry of Reward to Design Economic Auctions," *Science* 321, no. 5897 (2008): 1849–1852.

Chapter 23: Nature or Nurture

1. A. Knafo, S. Israel, A. Darvasi, R. Bachner-Melman, F. Uzefovsky, L. Cohen, E. Feldman, E. Lerer, E. Laiba, Y. Raz, L. Nemanov, I. Gritsenko, C. Dina, G. Agam, B. Dean, G. Bornstein, and R. P. Ebstein, "Individual Differences in Allocation of Funds in the Dictator Game Associated with Length of the Arginine Vasopressin 1a Receptor RS3 Promoter Region and Correlation Between RS3 Length and Hippocampal mRNA," *Gene and Brain Behavior* 7, no. 3 (2008): 266–275.

2. R. Ebstein, S. Israel, S. H. Chew, S. Zhong, and A. Knafo, "Genetics of Human Social Behavior," *Neuron* 65 (March 2010): 831–844.

INDEX

PROFESSOR EYAL WINTER is the director of the Center for the Study of Rationality at the Hebrew University of Jerusalem, one of the world's leading institutions in the academic study of decision making. Professor Winter served as chairman of the economics department at Hebrew University and was a recipient of the Humboldt Prize, awarded by the government of the Federal Republic of Germany, in 2011. He has also held positions as a professor at Washington University, the European University Institute in Florence, Italy, and the University of Manchester in the United Kingdom. In addition, Professor Winter is a council member of the International Game Theory Society and an associate editor of the journal *Games and Economic Behavior*. He has lectured at over 130 universities in 26 countries around the world, including Harvard University, Stanford University, Princeton University, the University of California at Berkeley, and the University of Cambridge.

PublicAffairs is a publishing house founded in 1997. It is a tribute to the standards, values, and flair of three persons who have served as mentors to countless reporters, writers, editors, and book people of all kinds, including me.

I. F. STONE, proprietor of *I. F. Stone's Weekly*, combined a commitment to the First Amendment with entrepreneurial zeal and reporting skill and became one of the great independent journalists in American history. At the age of eighty, Izzy published *The Trial of Socrates*, which was a national bestseller. He wrote the book after he taught himself ancient Greek.

BENJAMIN C. BRADLEE was for nearly thirty years the charismatic editorial leader of *The Washington Post*. It was Ben who gave the *Post* the range and courage to pursue such historic issues as Watergate. He supported his reporters with a tenacity that made them fearless and it is no accident that so many became authors of influential, best-selling books.

ROBERT L. BERNSTEIN, the chief executive of Random House for more than a quarter century, guided one of the nation's premier publishing houses. Bob was personally responsible for many books of political dissent and argument that challenged tyranny around the globe. He is also the founder and longtime chair of Human Rights Watch, one of the most respected human rights organizations in the world.

·   ·   ·

For fifty years, the banner of Public Affairs Press was carried by its owner Morris B. Schnapper, who published Gandhi, Nasser, Toynbee, Truman, and about 1,500 other authors. In 1983, Schnapper was described by *The Washington Post* as "a redoubtable gadfly." His legacy will endure in the books to come.

Peter Osnos, *Founder and Editor-at-Large*